W9-CSP-135

Introduction to Nonlinear Fluid-Plasma Waves

MECHANICS OF FLUIDS AND TRANSPORT PROCESSES
Editors: R. J. Moreau and G. Æ. Oravas

1. J. Happel and H. Brenner, Low Reynolds Number Hydrodynamics. 1983
   ISBN 90—247—2877—0
2. S. Zahorski, Mechanics of Viscoelastic Fluids. 1982
   ISBN 90—247—2687—5
3. J. A. Sparenberg, Elements of Hydrodynamic Propulsion. 1984
   ISBN 90—247—2871—1
4. B. K. Shivamoggi, Theoretical Fluid Dynamics. 1984
   ISBN 90—247—2999—8
5. R. Timman, A. J. Hermans and G. C. Hsiao, Water Waves and Ship Hydrodynamics. An Introduction. 1985
   ISBN 90—247—3218—2
6. M. Lesieur, Turbulence in Fluids. 1987
   ISBN 90—247—3470—3
7. L. A. Lliboutry, Very Slow Flows of Solids. 1987
   ISBN 90—247—3482—7
8. B. K. Shivamoggi, Introduction to Nonlinear Fluid-Plasma Waves. 1988
   ISBN 90—247—3662—5

Introduction to Nonlinear Fluid-Plasma Waves

By

Bhimsen K. Shivamoggi
University of Central Florida
Orlando, Fla., USA

Kluwer Academic Publishers
Dordrecht / Boston / London

Distributors

for the United States and Canada: Kluwer Academic Publishers, 101 Philip Drive, Norwell, MA 02061, USA
for the UK and Ireland: Academic Publishers, MTP Press Limited, Falcon House, Queen Square, Lancaster LA1 1RN, UK
for all other countries: Kluwer Academic Publishers Group, Distribution Center, P.O. Box 322, 3300 AH Dordrecht, The Netherlands

Library of Congress Cataloging in Publication Data CIP

Shivamoggi, Bhimsen K.
Introduction to nonlinear fluid-plasma waves.

(Mechanics of fluids and transport processes; 8)
Bibliography: p.
Includes index.
1. Plasma waves. 2. Nonlinear theories. I. Title.
II. Series: Mechanics of fluids and transport processes; v. 8.
QC718.S485 1988 530.4′4 87–34841

ISBN 90–247–3662–5

PRINTED IN THE NETHERLANDS

He is seen in Nature in the wonder of a flash of lightning.
Kena Upanishad (600 B.C.)

To My Parents

Contents

Preface

A variety of nonlinear effects occur in a plasma. First, there are the wave-steepening effects which can occur in any fluid in which the propagation speed depends upon the wave-amplitude. In a dispersive medium this can lead to classes of nonlinear waves which may have stationary solutions like solitons and shocks. Because the plasma also acts like an inherently nonlinear dielectric resonant interactions among waves lead to exchange of energy among them. Further, an electromagnetic wave interacting with a plasma may parametrically excite other waves in the plasma. A large-amplitude Langmuir wave undergoes a modulational instability which arises through local depressions in plasma density and the corresponding increases in the energy density of the wave electric field. Whereas a field collapse occurs in two and three dimensions, in a one-dimensional case, spatially localized stationary field structures called Langmuir solitons can result. Many other plasma waves like upper-hybrid waves, lower-hybrid waves etc. can also undergo a modulational instability and produce localized field structures. A new type of nonlinear effect comes into play when an electromagnetic wave propagating through a plasma is strong enough to drive the electrons to relativistic speeds. This leads to a propagation of an electromagnetic wave in a normally overdense plasma, and the coupling of the electromagnetic wave to a Langmuir wave in the plasma. The relativistic mass variation of the electrons moving in an intense electromagnetic wave can also lead to a modulational instability of the latter. Finally, a plasma has unique nonlinear properties which result from its basically discrete nature like those produced by the particles trapped in the electromagnetic potential of a wave propagating through the plasma.

Nonlinear effects on the propagation of waves in a plasma have been the object of intense research. These studies have also revealed that certain generic equations can be used to describe the nonlinear evolution of a large class of waves in a fluid plasma. The similarity of the asymptotic behavior of solutions to these generic equations suggests the possibility of a unified treatment of nonlinear phenomena and an apparent underlying "simplicity" in their description.

This book proposes to give an introductory account of theories of some macroscopic nonlinear phenomena with the fluid model for a plasma, with a further restriction to coherent processes. This book does not, of course, attempt to give a thorough and exhaustive coverage of this vast subject. As such, experts might notice omissions of some important items. Besides, in a rapidly evolving area like nonlinear plasma waves, a book can never get really "completed" but can merely escape from an author's hands at a chosen time. In the same vein, the list of references given at the end of this book is not to be construed as being complete. I have tried to make this book fairly self-contained, and an undergraduate level preparation in plasma physics is sufficient to follow material in this book, (a somewhat higher level mathematical preparation may be needed occasionally to handle a few items).

Orlando BHIMSEN K. SHIVAMOGGI

Acknowledgements

At the outset, I want to place it on record that without the help and encouragement of several individuals this book would not have become possible. My first gratitude is to Professor Mahinder Uberoi, who as my former teacher, guide and mountain-climbing partner has been a constant source of encouragement. Some of the time spent on this project has been supported by our Center for Research in Electro-Optics and Lasers and Professors Ron Phillips and Larry Andrews have given me considerable support and encouragement. Our Department Chairman Professor L. Debnath has provided me facilities in the preparation of this manuscript. My friend and senior colleague Professor Ram Mohapatra has given me much encouragement and advice on some mathematical aspects of the subject. I have had the benefit of many useful discussions with Professors Predhiman Kaw, Abhijit Sen and Ram Varma. The final version of the book-manuscript has benefitted considerably from the helpful suggestions of Professor Rene Moreau, the Series Editor, and the referees. The Publishers have also contributed to the success of this project through their excellent cooperation. My work on preparing material has been facilitated by the cheerful help given to me by Nancy Myler and Cheryl Mahan of our Inter-Library Loan Office in promptly processing my innumerable requests for procuring papers from elsewhere. My wife Jayashree has given me her gracious support in this undertaking. My parents, Pramila and Krishnarao, despite their limited education, have given me tremendous support in my endeavors to do advanced study and research (which is still a major luxury in the third world). Therefore, this book is dedicated to my parents as a token of my respect and gratitude.

BHIMSEN K. SHIVAMOGGI

I

Introduction

The linear wave theory is based on the small-amplitude approximation, wherein all nonlinear terms (i.e., terms quadratic or higher powers in the dependent variables and their derivatives are neglected). Consequently, it is possible to Fourier analyze any physical perturbation, and each harmonic can be treated independently. The propagation characteristics of the wave are usually expressed in the form of a dispersion relation $\omega = \omega(k)$, which relates the frequency and wavenumber of the wave in terms of plasma parameters. The waveform is purely sinusoidal (see Appendix).

The linear approximation can break down very easily. For example, in an experiment, if the wave is externally launched and is of large amplitude, the nonlinear terms can no longer be neglected. Similarly, if the wave is unstable, even in the linear sense, it quickly gets out of the linear regime since the predicted growth is exponential. In the nonlinear regime a variety of effects come into play and drastically change the linear picture of wave propagation.

First, there are all those nonlinear effects which can occur in any fluid in which the propagation speed depends on the wave amplitude. For example, an increase of wave velocity with amplitude leads to wave steepening (Dawson [1], Davidson and Schram [2], and Albritton and Rowlands [3]), and in a dispersive medium this can lead to classes of nonlinear waves which may have stationary solutions (example: ion-acoustic solitons, Washimi and Taniuti [4], or shocks, Sagdeev [5]). Solitons are of special interest because of the discovery of Zabusky and Kruskal [6] that they preserve their identity and are stable in mutual collisions. An ingenius technique to construct an exact solution of the initial-value problem of the Korteweg—de Vries equation which has solitons as its solutions was given by Gardner *et al.* [7]. Experimental observations of ion-acoustic solitons were made by Ikezi *et al.* [8, 9], and those of ion-acoustic shocks were by Alikhanov *et al.* [10], and Taylor *et al.* [11].

In addition, because the plasma also acts as an inherently nonlinear dielectric medium, sets of waves which satisfy resonance criteria

$$\sum_i \mathbf{k}_i = \mathbf{o}, \qquad \sum_i \omega_i(\mathbf{k}_i) = 0$$

are nonlinearly coupled so that energy can be transferred to different regions of **k**-space, for example, from initially unstable wavelengths to dissipating wavelengths. Such an interaction may constitute decay of a pump wave into plasma waves (Silin [12], DuBois and Goldman [13]). Nonlinear resonant interaction between two electromagnetic waves and a Langmuir wave was treated by Sjolund and Stenflo [14] and Shivamoggi [15]. Nonlinear resonant interaction between two electromagnetic waves and an ion-acoustic wave was treated by Lashmore-Davies [16] and Shivamoggi [15]. Nonlinear resonant interaction between two circularly-polarized waves and a Langmuir wave all propagating parallel to the direction of the applied magnetic field was considered by Sjolund and Stenflo [17], Chen and Lewak [18], Prasad [19], Lee [20], and Shivamoggi [21]. Nonlinear resonant interaction between three ordinary electromagnetic waves propagating perpendicular to the applied magnetic field was treated by Stenflo [22, 23], Krishan *et al.* [24], and Munoz and Dagach [25]. Nonlinear resonant interaction between three extraordinary electro-magnetic waves propagating perpendicular to the applied magnetic field was considered by Harker and Crawford [26], Das [27] and Shivamoggi [21]. Experimental observations of wave-wave interactions were made by Porkolab and Chang [28], Phelps *et al.* [29], and Franklin *et al.* [30].

On the other hand, if such a resonant process results from an applied monochromatic high-frequency field, then it is called a parametric process (Montgomery and Alexeff [31], and Nishikawa [32]). One of the consequences of the interaction of an incident (pump) electromagnetic wave with a plasma is the parametric excitation of two plasma waves. If the latter are both purely electrostatic, they are eventually absorbed in the plasma, and this decay process then leads to enhanced (or anomalous) absorption of the incident electromagnetic wave. If one of the excited plasma waves is electromagnetic, it can escape from the plasma and show up as enhanced (or stimulated) scattering of the incident electro-magnetic wave. The latter process can be of two types according as the other excited plasma wave is a Langmuir wave (stimulated Raman scattering) or an ion-acoustic wave (stimulated Brillouin scattering), (Bornatici [33], and Shivamoggi [34]). The presence of stimulated Raman scattering and stimulated Brillouin scattering has been confirmed in several experiments (Ripin *et al.* [35], Offenberger *et al.* [36, 37], Turechek and Chen [38], Grek *et al.* [39], Watt *et al.* [40], Ng *et al.* [41]).

In a plasma acted on by an intense laser field, one of the products is a spectrum of excited Langmuir waves. As the amplitudes of the latter build up they too can undergo further decay instabilities with the consequent excitation of Langmuir waves of smaller wavenumbers and lower frequencies (Valeo *et al.* [42], DuBois and Goldman [43]). For a hot, tenuous plasma collisions between the particles will be negligible, and as the wavenumber of the Langmuir waves decreases the effect of Landau damping will also become negligible. A problem then arises as to how this large amount of energy concentrated in the low wavenumber end of the Langmuir spectrum will be dissipated. Zakharov [44, 45] proposed that this spectral concentration of energy would induce a nonlinear modulational instability of Langmuir waves which arises through local depressions in density and the corresponding increases in the energy density of the Langmuir electric field. The density depressions are produced by the low-frequency ponderomotive force due to the beating of the Langmuir waves with each other on electrons, expelling them (and hence the ions as well through ambipolar effects) from the region of strong fields. Such density cavities act as potential wells and trap the Langmuir electric field and intensify them. As the electric field grows, more plasma is removed and the field is trapped and intensified further until a collapse of the field may occur in two and three-dimensional cases — this result has been confirmed by the 'virial theorem' for the time evolution of the wave-packet averaged width of an initial wave packet in the adiabatic approximation (Goldman and Nicholson [46]), and the numerical solutions of an initial-value problem associated with Zakharov's equations (Pereira *et al.* [47], Nicholson *et al.* [48], Hafizi *et al.* [49]). This spatial collapse is believed to be a consequence of the instability of the localized solutions of Zakharov's equations to two-dimensional perturbations (Schmidt [50]). When the collapsed states attain spatial scales of the order of the Debye length, Landau damping sets in and the wave energy is then dissipated into particle energy. In a one-dimensional case, however, it is possible that the rate at which the field energy is trapped within the cavity can be balanced by the rate at which the field energy can leak out due to convection. Under such circumstances, a spatially localized stationary field structure called Langmuir soliton can result, (Karpman [51, 52], Rudakov [53]).

Mathematically, the Langmuir solitons are localized stationary solutions of a nonlinear Schrödinger equation for the Langmuir field with an effective potential proportional to the low-frequency electron-density perturbation. The latter is in turn governed by an equation for ion-acoustic waves driven by the ponderomotive force associated with the Langmuir field. In the adiabatic limit, one obtains the cubic nonlinear Schrödinger equation. Zakharov and Shabat [54] used the inverse-scattering formalism to solve this equation exactly for initial conditions which represent localized states. Another property exhibited by the solutions of the cubic nonlinear Schrödinger equation is that of recurrence of states, which was found numerically by Morales *et al.* [55] during a study of the long-time evolution of these solutions. Thyagaraja [56, 57] showed that the existence of integral invariants (Gibbons *et al.* [58]) of the cubic nonlinear Schrödinger equation plays a vital role in establishing the recurrence properties.

The envelope soliton solutions of Zakharov's equations are not valid for propagation close to the speed of sound. In order to develop a solution valid for the latter case, one has to include the nonlinearities in the ion motion (Makhankov [59], Nishikawa *et al.* [60], and Rao and Varma [61]). Rao and Varma [61] gave a solution which showed that a single-humped soliton becomes a double-humped one as the speed of propagation increases. This problem can be treated by a simpler method like the one given in Shivamoggi [62].

In all of the above treatments of the problem of nonlinearly-coupled Langmuir and ion-acoustic waves, the Langmuir wave has been treated as the driving wave, and as a consequence, one has picked out for investigation a somewhat specialized subclass of solutions. A generalized perturbation theory has been given by Shivamoggi [63] which reveals the nature of coupled oscillatory waves in the system.

Experimental observations of self-focusing and modulational instability of a laser beam were made by Garmire *et al.* [64], and Campillo *et al.* [65]. Observations of density cavities and associated localized Langmuir wave envelopes were made by Ikezi *et al.* [66, 67], Kim *et al.* [68], Cheung *et al.* [69], Wong and Quon [70], and Leung *et al.* [71].

A model for strong Langmuir turbulence was formulated by Kingsep *et al.* [72] which consisted not of a Fourier superposition of plane waves, but of a collection of one-dimensional Langmuir solitons. Galeev *et al.* [73, 74] gave a more accurate model of strong Langmuir turbulence, valid in three dimensions, and based on an 'ideal gas' of self-similarly collapsing solitons. Experimental observation of such a model of strong Langmuir turbulence seems to have been made by Wong and Cheung [75]. Excellent theoretical accounts of strong Langmuir turbulence have been given by Rudakov and Tsytovich [76], and Thornhill and ter Haar [77].

Modulational stability of ion-acoustic waves was considered by Shimizu and Ichikawa [78], Kakutani and Sugimoto [79], Chan and Seshadri [80], Infeld and Rowlands [81], and Shivamoggi [82]. Shimuzu and Ichikawa [78] used the reductive perturbation method, and Kakutani and Sugimoto [79], Chan and Seshadri [80] used the Krylov-Bogoliubov-Mitropolski method to derive a nonlinear Schrödinger equation which predicted the ion-acoustic waves to be stable to modulations of long wavelength. Shivamoggi [82] derived a different nonlinear Schrödinger equation for the ion-acoustic waves instead by using the Zakharov-Karpman approach wherein the ion-acoustic wave interacts with a slow plasma motion, and obtained the same result. Ikezi *et al.* [83] made an experimental verification of these results.

Modulational instability and formation of envelope solitons of the upper-hybrid wave have been considered by Kaufmann and Stenflo [84], Porkolab and Goldman [85], and Shivamoggi [62, 86]. A strong magnetic field was found to eliminate the existence of subsonic upper-hybrid solitons but to make possible the existence of supersonic upper-hybrid solitons. In order to treat the upper-hybrid solitons travelling at speeds near the speed of sound the ion-nonlinearity has to be taken into account (Shivamoggi [62]). Shivamoggi [86] established the existence of upper-hybrid waves in an electron plasma co-existing with a uniform immobile positive ion background by explicitly evaluating the amplitude-dependent nonlinear frequency-shift for the upper-hybrid wave in a cold plasma and by introducing then the effects due to the thermal dispersion in the warm plasma. Since the ion motions are neglected, Zakharov's mechanism for the production of upper-hybrid solitons (Shivamoggi [62]) would predict the latter to exist only at supersonic speeds or when the thermal dispersion is negative. The latter was indeed found (Shivamoggi [86]) to be the condition for the existence of upper-hybrid solitons in an electron plasma! Experimental observations of upper-hybrid solitons were made by Cho and Tanaka [87].

Experiments of Gekelman and Stenzel [88] showed that the lower-hybrid waves were localized in a spatial wave packet propagating into the plasma along a conical trajectory which makes a small angle with respect to the confining magnetic field. Morales and Lee [89] traced these conical wave packets to the filamentation through self-interaction of a large-amplitude lower-hybrid wave in the plasma. This was suggested to occur through Zakharov's mechanism wherein density cavities were created by the ponderomotive force exerted by the lower-hybrid waves on both the electrons and ions which in turn trap the latter.

The self-modulation of electromagnetic whistler waves in magnetized plasmas were considered by Tam [90], Taniuti and Washimi [91], Hasegawa [92, 93], and Washimi and Karpman [94, 95]. The nonlinear modulation of Alfvén waves was considered by Rogister [96], Mjolhus [97, 98], Spangler and Sheerin [99], and Lashmore-Davies and Stenflo [100].

A new type of nonlinear effects comes into play when an electromagnetic wave propagating through a plasma is strong enough to drive the electrons to relativistic speeds, (Akhiezer and Polovin [101], Lunow [102], Kaw and Dawson [103], Max and Perkins [104], Chian and Clemmow [105], DeCoster [106], Kaw *et al.* [107], and Shivamoggi [63]). Two nonlinear effects which arise in this context are:

(i) relativistic variation of the electron mass;
(ii) excitation of space-charge fields by strong $\mathbf{V} \times \mathbf{B}$ forces driving electrons along the direction of propagation of the electromagnetic wave.

The first effect leads to a propagation of the electromagnetic wave in a normally overdense plasma (recall that according to linear theory of propagation of electromagnetic waves in a plasma, an electromagnetic wave with a frequency less than the plasma frequency cannot propagate in the plasma). The second effect leads to a coupling of the electro-magnetic wave to the Langmuir wave. All of the treatments given in [101]—[105] were concerned with obtaining exact (or nearly exact) analytical solutions for some special cases like purely transverse waves and purely longitudinal waves. DeCoster [106] considered the general case, for which he gave some approximate solutions. However, DeCoster's procedure does not allow for amplitude modulations of the coupled waves, and is, therefore, restricted in scope. Actually, DeCoster's calculations will be invalid, if the system of coupled waves in question exhibits internal resonances with the concomitant modulations in the amplitudes of the coupled waves. Shivamoggi [63] gave a generalized perturbation theory that allows for both amplitude — and phase — modulations of the coupled waves. This theory can, therefore, successfully deal with internal resonances if they arise in the system in question. This theory also recovered results of the known special cases like the relativistic nonlinear frequency shift of an electromagnetic wave given by Sluijter and Montgomery [108] and that of a longitudinal wave given by Akhiezer and Polovin [101] in the appropriate limit.

Experiments of Stamper *et al.* [109—111] and Diverglio *et al.* [112] showed that intense spontaneously generated magnetic fields are present in laser-produced plasmas. The nonlinear relativistic propagation of electromagnetic waves in a magnetized plasma was, therefore, considered by Akhiezer and

Polovin [101], Aliev and Kuznetsov [113], Berzhiani *et al.* [114] and Shivamoggi [115] who considered in particular, the nonlinear propagation of an intense circularly-polarized electromagnetic wave along a constant applied magnetic field. Shivamoggi [115] showed that the incident circularly-polarized electromagnetic wave propagates farther into denser regions of the plasma in the presence of an applied magnetic field.

The relativistic mass variation of the electrons moving in an intense electromagnetic wave can also lead to a modulational instability of the latter (Max *et al.* [116]). The relativistic modulational instability of an intense electromagnetic wave was also considered by Shivamoggi [117], who gave a new method for investigation of this problem that involved the derivation of a nonlinear Schrödinger equation.

It may be mentioned, though we do not consider them in this book that a plasma has unique nonlinear properties which result from its basically discrete nature like those produced by the particles trapped in the electromagnetic potential of a wave propagating through the plasma. Indeed, the ever-present discrete-particle effects in a laboratory plasma may sometimes obscure the fluid phenomena! An excellent summary of the discrete particle effects has been given recently by Pecseli (Pecseli, H.: IEEE Trans. Plasma Sci. *PS-13*, 53 (1985)).

II

Nonlinear Oscillations in an Electron Plasma

Consider the simple longitudinal oscillations of the electron in a plasma: plasma oscillations, which can be modelled by a one-dimensional cold plasma (since $\omega/k \gg V_{T_e}$, V_{T_e} being the thermal speed of the electrons) in the fluid approximation. For the high-frequency plasma waves, the ions can be assumed to be immobile and to provide a uniform background of positive charge. The one-dimensional electrostatic motions of the electrons are then governed by the following equations

$$\frac{\partial n}{\partial t} + \frac{\partial}{\partial x}(nv) = 0 \tag{1}$$

$$\frac{\partial V}{\partial t} + V\frac{\partial V}{\partial x} = -\frac{e}{m}E \tag{2}$$

$$\frac{\partial E}{\partial x} = -4\pi e(n - N_0) \tag{3}$$

$$\frac{\partial E}{\partial t} - 4\pi env = 0 \tag{4}$$

where V is the plasma velocity in the x-direction, E is the electrostatic field, n is the number density of the electrons, N_0 is the unperturbed value of n, and all other quantities have their usual meaning.

In the Eulerian description as in equations (1)—(4), one pays attention to only that fluid particle that happens to be at (x, t). However, in a Lagrangian description, one keeps track of a particular fluid particle (x_0, τ). The discovery of the merits of a Lagrangian description for this problem is due to Dawson [1]. The relation between the two descriptions is

$$x = x_0 + \int_0^\tau V_0(x_0, \tau')\, d\tau', \quad \tau = t \tag{5}$$

from which,

$$\left(\frac{\partial x_0}{\partial x}\right)_\tau = \frac{1}{1 + \int_0^\tau \frac{\partial V_0}{\partial x_0}\, d\tau'} \tag{6}$$

and

$$\left(\frac{\partial x_0}{\partial \tau}\right)_x = \frac{-V_0(x_0, \tau)}{1 + \int_0^\tau \frac{\partial V_0}{\partial x_0}\, d\tau'}. \tag{7}$$

Let us denote

$$G(x, t) \equiv G_0(x_0, \tau). \tag{8}$$

Note,

$$\begin{aligned} \left(\frac{\partial G}{\partial x}\right)_t &= \left(\frac{\partial G}{\partial x_0}\right)_\tau \left(\frac{\partial x_0}{\partial x}\right)_t \\ \left(\frac{\partial G}{\partial t}\right)_x &= \left(\frac{\partial G_0}{\partial \tau}\right)_{x_0} + \left(\frac{\partial G_0}{\partial x_0}\right)_\tau \left(\frac{\partial x_0}{\partial \tau}\right)_x. \end{aligned} \tag{9}$$

Using (6)—(9), equations (2) and (4) become

$$\frac{\partial V_0}{\partial \tau} = -\frac{e}{m} E_0 \tag{10}$$

$$\frac{\partial E_0}{\partial \tau} = 4\pi e N_0 V_0 \tag{11}$$

from which,

$$\frac{\partial^2 V_0}{\partial \tau^2} + \omega_p^2 V_0 = 0 \tag{12}$$

where,

$$\omega_p^2 \equiv \frac{4\pi N_0 e^2}{m}.$$

Therefore,

$$V_0(x_0, \tau) = V(x_0) \cos \omega_p \tau + \omega_p X(x_0) \sin \omega_p \tau \tag{13}$$

and

$$E_0(x_0, \tau) = \frac{m}{e} \omega_p [V(x_0) \sin \omega_p \tau - \omega_p X(x_0) \cos \omega_p \tau]. \tag{14}$$

Thus, in the Lagrangian frame, an initial excitation oscillates coherently and indefinitely at the electron-plasma frequency in a sinusoidal fashion.

Using (6)—(9), equation (1) becomes

$$\frac{\partial n_0}{\partial \tau} + n_0 \frac{\partial V_0}{\partial x_0} \frac{1}{1 + \int_0^\tau \frac{\partial V_0}{\partial x_0} \, \mathrm{d}\tau'} = 0$$

or

$$\frac{\partial}{\partial \tau}\left[n_0 \left(1 + \int_0^\tau \frac{\partial V_0}{\partial x_0} \, \mathrm{d}\tau' \right) \right] = 0$$

from which,

$$n_0(x_0, \tau) = \frac{n_0(x_0, 0)}{1 + \int_0^\tau \frac{\partial V_0}{\partial x_0} \, \mathrm{d}\tau'} . \tag{15}$$

Using (13), we obtain

$$n_0(x_0, \tau) = \frac{n_0(x_0, 0)}{\left[1 + \frac{1}{\omega_p} \frac{\mathrm{d}V}{\mathrm{d}x_0} \sin \omega_p \tau + \frac{\mathrm{d}X}{\mathrm{d}x_0} (1 - \cos \omega_p \tau) \right]} . \tag{16}$$

From (5) and (13), one obtains

$$x = x_0 + \frac{V(x_0)}{\omega_p} \sin \omega_p \tau + X(x_0)(1 - \cos \omega_p \tau). \tag{17}$$

From (3), (14) and (16), one also has

$$\frac{\mathrm{d}X}{\mathrm{d}x_0} = \frac{n_0(x_0, 0)}{N_0} - 1. \tag{18}$$

In order to see what happens in the laboratory frame, one has to transform back to Eulerian variables (x, t) using (5). For this it is necessary to prescribe the initial conditions. As a simple example, let us choose

$$\begin{aligned} n_0(x_0, 0) &= N_0(1 + \Delta \cos kx_0) \\ V(x_0) &= 0 \end{aligned} \tag{19}$$

following Davidson and Schram [2].

Equations (13), (14) and (16) then become

$$V_0(x_0, \tau) = \frac{\omega_p}{k} \Delta \sin kx_0 \cdot \sin \omega_p \tau$$

$$E_0(x_0, \tau) = -\frac{m}{e} \omega_p^2 \frac{\Delta}{k} \sin kx_0 \cdot \cos \omega_p \tau \tag{20}$$

$$n_0(x_0, \tau) = N_0 \frac{1 + \Delta \cos kx_0}{1 + \Delta \cos kx_0 \cdot (1 - \cos \omega_p \tau)}$$

with

$$kx = kx_0 + \Omega(\tau) \sin kx_0, \quad \tau = t$$
$$\Omega(\tau) \equiv 2\Delta \sin^2 \frac{\omega_p \tau}{2} . \tag{21}$$

One requires $\Delta < \frac{1}{2}$, so $|\Omega(\tau)| < 1$, and then $x_0(x, t)$ will be single-valued.

If one writes

$$\sin kx_0 = \sum_{n=1}^{\infty} a_n(t) \sin nkx \tag{22}$$

where

$$a_n(t) = \frac{k}{\pi} \int_0^{2\pi/k} \sin nkx \cdot \sin kx_0 \cdot \mathrm{d}x,$$

using (21), one then obtains

$$a_n(t) = (-1)^{n+1} \frac{2}{n\Omega(t)} J_n[n\Omega(t)] \tag{23}$$

where $J_n(z)$ is the Bessel function of nth order.

Using (22) and (23), (20) becomes

$$V(x, t) = \frac{\omega_p}{k} \Delta \sum_{n=1}^{\infty} (-1)^{n+1} \frac{2}{n\Omega(t)} J_n[n\Omega(t)] \times$$
$$\times \sin nkx \cdot \sin \omega_p t \tag{24}$$

$$E(x, t) = -\frac{m}{e} \frac{\omega_p^2}{k} \Delta \sum_{n=1}^{\infty} (-1)^{n+1} \frac{2}{n\Omega(t)} J_n[n\Omega(t)] \times$$
$$\times \sin nkx \cdot \cos \omega_p t \tag{25}$$

$$n(x, t) = N_0 \left\{ 1 + \frac{2\Delta}{\Omega(t)} \sum_{n=1}^{\infty} (-1)^{n+1} J_n[n\Omega(t)] \times \right.$$

$$\left. \times \cos nkx \cdot \cos \omega_p t \right\} \tag{26}$$

where $J_n(x)$ is Bessel's function of nth order.

Equations (24)—(26) are exact solutions to the initial-value problem. Note that the wave form in the laboratory frame is no longer sinusoidal but is distorted through the generation of higher harmonies of kx. In particular, notice the nonlinear-steepening effect (Figure 2.1). Another interesting feature in (24)—(26) is the additional time dependence (apart from $\omega_p t$) which enters the argument of the Bessel functions, and corresponds to a nonlinear frequency shift of the oscillation. It turns out that for a particular choice of initial conditions, one may obtain the travelling waves from the above solutions. Following Albritton and Rowlands [3], let us choose

$$\begin{aligned} V(x_0) &= A\omega_p \cos ky_0 \\ X(x_0) &= A \sin ky_0 \end{aligned} \tag{27}$$

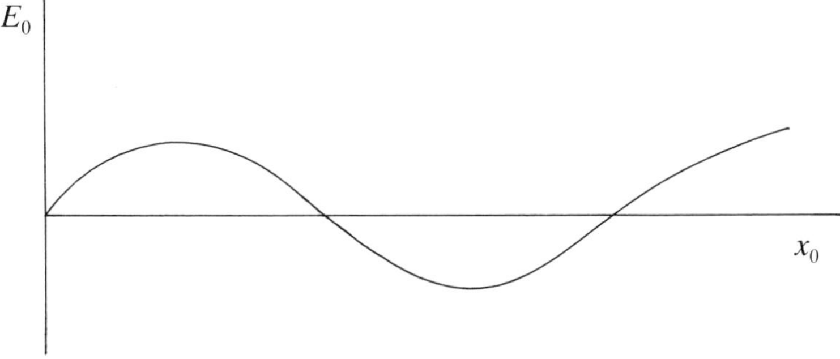

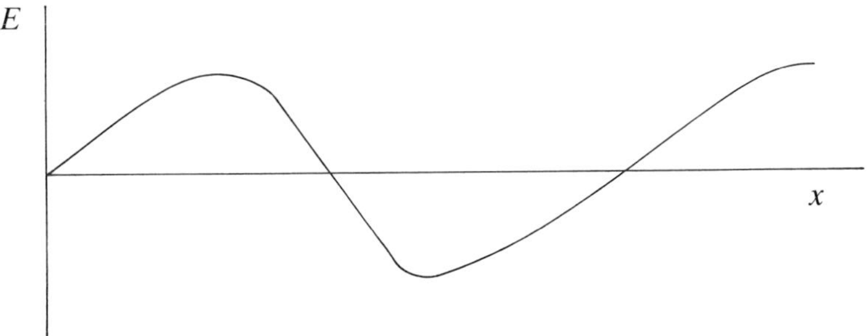

Figure 2.1. Nonlinear steepening effect.

where

$$y_0 = x_0 + X(x_0)$$

then (14) becomes

$$E_0(x_0, \tau) = -\frac{m}{e}\,\omega_p^2 A\,\sin(\omega_p \tau - k y_0) \tag{28}$$

where

$$x = y_0 + A\,\sin(\omega_p t - k y_0). \tag{29}$$

Inverting (29), as before, (28) becomes

$$E(x, t) = 2\,\frac{m}{e}\,\frac{\omega_p^2}{k^2}\sum_{n=1}^{\infty}\frac{J_n(nkA)}{n}\,\sin[n(kx - \omega_p t)] \tag{30}$$

which corresponds to a travelling wave moving with a phase velocity ω_p/k.

Finally, let us briefly examine the effects of including thermal motion of the electrons (Coffey [118]). This can be done simply by introducing an additional term

$$-\frac{1}{nm}\,\frac{\partial p}{\partial x}$$

on the right hand side of equation (2). Then, to close the set of equations, we need one more equation — the equation of state:

$$\frac{p}{n^3} = \text{const.} \tag{31}$$

which is valid for $\omega/k \gg V_{T_e}$, V_{T_e} being the thermal speed of electrons.

Transforming to Lagrangian variables, as before, one obtains in place of equation (12):

$$\frac{\partial^2 V_0}{\partial \tau^2} + \omega_p^2 V_0 = \frac{3}{mn(x, 0)}\,\frac{\partial}{\partial x_0}\left[\frac{p_0(x_0, 0)\,\dfrac{\partial V_0}{\partial x_0}}{\left\{1 + \displaystyle\int_0^{\tau}\frac{\partial V_0}{\partial x_0}\,\mathrm{d}\tau'\right\}^4}\right] \tag{32}$$

which, unlike the cold-plasma case, is now a nonlinear equation and hence not easily tractable. If $k^2 V_{t_e}^2/\omega_p^2 \ll 1$, then one can treat the thermal contribution as a small correction and linearize equation (32) (for $p_0(x_0, 0) = P$, $n_0(x_0, 0) = N$ and $E_0(x_0, 0) = 0$) to get

$$\frac{\partial^2 V_0}{\partial \tau^2} + \omega_p^2 V_0 = 3\omega_p^2 \lambda_D^2\,\frac{\partial^2 V_0}{\partial x_0^2} \tag{33}$$

where

$$\lambda_D^2 \equiv \frac{V_{T_e}^2}{\omega_p^2}$$

the solution of equation (33) for initial periodic perturbations is again similar to those obtained earlier for the cold-plasma case, i.e., temporal oscillations that persist indefinitely and have period $2\pi/\omega(k)$. The only modification in this case is that there is a small frequency shift from the cold plasma result, since

$$\omega(k) = \omega_p[1 + 3k^2\lambda_D^2]^{1/2}. \tag{34}$$

However, if the initial perturbation is localised, the dispersive effects of electron thermal motion cause the disturbance to spread in space indefinitely with

$$\operatorname*{Lt}_{\tau \Rightarrow \infty} V_0(x_0, \tau) = 0.$$

This is in contrast to the cold-plasma case, where coherent oscillations are maintained indefinitely over the region of initial excitation. Physically this is due to the fact that the phase velocity ω/k of the wave is now a function of the wavenumber. Therefore, for a localised initial perturbation, its various Fourier components will travel at different velocities, causing it to spread. In the laboratory frame this tendency will appear together with the wave-steepening effect associated with harmonic generation, and the two effects oppose each other. Interesting situations arise when the two effects are comparable, as we will see in the next chapter. However, in a warm electron plasma, as we will see below, a balance cannot be struck between the above two opposing effects, and a progressive wave-steepening appears to be the only way in which a nonlinear wave in a warm electron plasma can evolve.

Consider a one-dimensional wave motion in a warm isothermal electron plasma, with the ions merely forming an immobile neutralizing background. The equations governing such a wavemotion are:

$$\frac{\partial n}{\partial t} + \frac{\partial}{\partial x}(nv) = 0 \tag{35}$$

$$\frac{\partial v}{\partial t} + v\frac{\partial v}{\partial x} = -\frac{1}{n}\frac{\partial n}{\partial x} + \frac{\partial \phi}{\partial x} \tag{36}$$

$$\frac{\partial^2 \phi}{\partial x^2} = n - 1 \tag{37}$$

where n is the electron number density normalized by the mean value n_0, v is the electron fluid velocity non-dimensionalized by the electron thermal speed $V_{T_e} = \sqrt{KT_e/m}$, ϕ is the electric potential normalized by KT_e/e, x by the Debye length $\lambda_e = V_{T_e}/\omega_{p_e}$ and t by $\omega_{p_e}^{-1}$.

Let us now consider a slowly-varying wavetrain in a warm electron plasma of the form:

$$\begin{aligned} n(x, t) &= n(\xi, \tau; \varepsilon) \\ v(x, t) &= v(\xi, \tau; \varepsilon) \\ \phi(x, t) &= \phi(\xi, \tau; \varepsilon) \end{aligned} \tag{38}$$

where $\varepsilon \ll 1$ is a small parameter which may characterize a typical wave-amplitude, and

$$\xi \equiv kx - \omega t, \qquad \tau \equiv \varepsilon t. \tag{39}$$

Equations (35)—(37), then become

$$\varepsilon \frac{\partial n}{\partial \tau} - \omega \frac{\partial n}{\partial \xi} + k \frac{\partial}{\partial \xi}(nv) = 0 \tag{40}$$

$$\varepsilon \frac{\partial v}{\partial \tau} - \omega \frac{\partial v}{\partial \xi} + kv \frac{\partial v}{\partial \xi} = -\frac{k}{n} \frac{\partial n}{\partial \xi} + k \frac{\partial \phi}{\partial \xi} \tag{41}$$

$$k^2 \frac{\partial^2 \phi}{\partial \xi^2} = n - 1. \tag{42}$$

Let us look for solutions of the form

$$\begin{aligned} n &= 1 + \varepsilon n_1 + \varepsilon^2 n_2 + \dots \\ v &= \varepsilon v_1 + \varepsilon^2 v_2 + \dots \\ \phi &= \varepsilon \phi_1 + \varepsilon^2 \phi_2 + \dots \end{aligned} \tag{43}$$

Equations (40)—(42), then give:

$0(\varepsilon)$:

$$-\omega \frac{\partial n_1}{\partial \xi} + k \frac{\partial v_1}{\partial \xi} = 0 \tag{44}$$

$$-\omega \frac{\partial v_1}{\partial \xi} + k \frac{\partial n_1}{\partial \xi} - k \frac{\partial \phi_1}{\partial \xi} = 0 \tag{45}$$

$$n_1 - k^2 \frac{\partial^2 \phi_1}{\partial \xi^2} = 0 \tag{46}$$

$0(\varepsilon^2)$:

$$-\omega \frac{\partial n_2}{\partial \xi} + k \frac{\partial v_2}{\partial \xi} = -k \frac{\partial}{\partial \xi}(n_1 v_1) - \frac{\partial n_1}{\partial \tau} \tag{47}$$

$$-\omega \frac{\partial v_2}{\partial \xi} + k \frac{\partial n_2}{\partial \xi} - k \frac{\partial \phi_2}{\partial \xi} = -kv_1 \frac{\partial v_1}{\partial \xi} + kn_1 \frac{\partial n_1}{\partial \xi} - \\ - \frac{\partial v_1}{\partial \tau} \tag{48}$$

$$n_2 - k^2 \frac{\partial^2 \phi_2}{\partial \xi^2} = 0. \tag{49}$$

Equations (44)—(46) give the linear result:

$$n_1 = \frac{k}{\omega} v_1 = k^2 \frac{\partial^2 \phi_1}{\partial \xi^2}$$
$$\phi_1 = A(\tau) \cos \xi, \quad \omega^2 = 1 + k^2. \tag{50}$$

Using (50), equations (47)—(49) give:

$$k(k^2 - \omega^2) \frac{\partial^3 \phi_2}{\partial \xi^3} - k \frac{\partial \phi_2}{\partial \xi} = -2k\omega \frac{\partial^3 \phi_1}{\partial \xi^2 \partial \tau} + $$
$$+ k^3 \left[1 - \omega^2 \left(1 + \frac{2}{k^2} \right) \right] \frac{\partial^2 \phi_1}{\partial \xi^2} \frac{\partial^3 \phi_1}{\partial \xi^3} . \tag{51}$$

The removal of secular terms in equation (51) requires:

$$\frac{\partial^3 \phi_1}{\partial \xi^2 \partial \tau} = 0 \quad \text{or} \quad A \neq A(\tau). \tag{52}$$

The solution of equation (51) is then given by

$$\phi_2 = \frac{A^2}{12} (k^4 + 2k^2 + 2) \sin 2\xi. \tag{53}$$

Equations (50), (52) and (53) show that the system (35)—(37) possesses only purely-periodic-wave solutions, and thus it is impossible for a nonlinear warm electron-plasma wave to evolve into a Korteweg—de Vries type solitary wave.

Nonlinear Ion-Acoustic Waves

Consider a plasma consisting of cold ions and hot electrons. Let us consider low-frequency waves in such a plasma. The electrons will then remain in thermodynamic equilibrium at a constant temperature T_e in these wave motions with a number density given by the Boltzmann distribution

$$n_e = n_0 e^{e\phi/KT_e} \tag{1}$$

where ϕ is the electrostatic potential associated with the wavemotions, and n_0 is the number density of electrons (or ions) in the unperturbed state. For low frequency waves we are considering, the underlying motion is that of the ions, with the electrons merely forming a screening cloud about the ions. Usually, the charge separation due to the ion-motion is immediately screened by the electrons so that the creation of ion-plasma waves is counteracted by the electrons. However, if the electron thermal speed is large enough, the electrons will not completely shield all electrostatic fields.

(1) Shock Waves

Consider a collisionless, one-dimensional shock wave which develops from a large-amplitude ion wave. Let the wave travel to the left with a velocity V_0. If one goes to the frame moving with the wave, the electrostatic potential $\phi(x)$ will be constant in time and one will see a stream of plasma impinging on the wave from the left with a velocity V_0. Since $T_i = 0$, all the ions are incident with the same velocity V_0. The electrons are Maxwellian, and if the shock wave moves much more slowly than the electron thermal speed, the shift in the center velocity of the Maxwellian can be neglected. The velocity of the ions in the shock wave is, from energy conservation,

$$V = \left(V_0^2 - \frac{2e\phi}{m_i} \right)^{1/2}. \tag{2}$$

The number density of the ions in the shock is then given by

$$n_i = \frac{n_0 V_0}{V} = \frac{n_0}{\left(1 - \dfrac{2e\phi}{m_i V_0^2}\right)^{1/2}}. \tag{3}$$

Poisson's equation then becomes

$$\begin{aligned}\frac{d^2\phi}{dx^2} &= 4\pi e(n_e - n_i)\\ &= 4\pi n_0 e\left[e^{e\phi/KT_e} - \left(1 - \frac{2e\phi}{m_i V_0^2}\right)^{-1/2}\right].\end{aligned} \tag{4}$$

Put

$$\chi = \frac{e\phi}{KT_e}, \quad \xi = \frac{x}{\lambda_D}, \quad M = \frac{V_0}{C_s}, \quad C_s^2 = \frac{KT_e}{m_i} \tag{5}$$

so that equation (4) becomes

$$\frac{d^2\chi}{d\xi^2} = e^{\chi} - \left(1 - \frac{2\chi}{M^2}\right)^{-1/2} \equiv -\frac{dV(\chi)}{d\chi}. \tag{6}$$

Using

$$\chi = 0: \quad V(\chi) = 0 \tag{7}$$

one obtains

$$V(\chi) = 1 - e^{\chi} + M^2\left[1 - \left(1 - \frac{2\chi}{M^2}\right)^{1/2}\right] \tag{8}$$

which is sketched in Figure 3.1.

If $V(\chi)$ has a well shape as shown in Figure 3.1, the quasiparticle described by χ will make a single excursion to $\chi > 0$ and return to $\chi = 0$. Such a pulse is called a solitary wave. In order that a well shape exists for $V(\chi)$, one requires $V(\chi) < 0$ for small χ, or

$$-\frac{\chi^2}{2} + \frac{\chi^2}{2M^2} < 0$$

or

$$M^2 > 1.$$

One further requires that $V(\chi) > 0$ for some $\chi > 0$. Noting that $\chi \leqslant M^2/2$, this gives

$$e^{M^2/2} - 1 < M^2$$

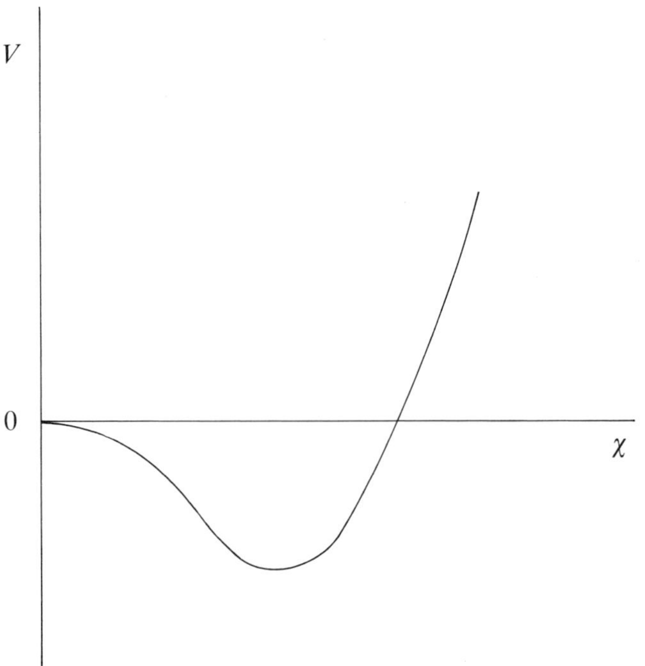

Figure 3.1. The quasi-potential well.

or

$$M^2 < 1.6.$$

Thus such waves exist only for a certain range of M, (Sagdeev [5]), *viz.*,

$$1 < M^2 < 1.6. \tag{9}$$

In the presence of collisions between ions and electrons, there will be energy loss and the solution will, therefore, never return to $\chi = 0$, but will eventually settle at the minimum of the potential $V(\chi)$. Such a structure would, indeed, represent a shock (Figure 3.2).

Observations of collisionless electrostatic ion-acoustic shocks were made by Alikhanov *et al.* [10] and by Taylor *et al.* [11]. In the experiment of Taylor *et al.* [11], large-amplitude ion-acoustic waves were excited by the interpenetration of two plasmas with high electron-to-ion temperature ratios ($6 < T_e/T_i < 20$). A compressional wave with a ramp shape is found to steepen in a fashion consistent with the classical overtaking discussed in Chapter II. This steepening continued until dispersive short scale oscillations developed at the front. Ions streaming in front of the shock were also observed. The number of ions streaming in front of the shock was found to depend on the excited wave potential and the ratio T_e/T_i. Figure 3.3 shows the change in shock propagation when T_e/T_i is changed by a factor of two. The number of streaming ions is seen to increase as T_e decreases. The damping of oscillations is due to collisions.

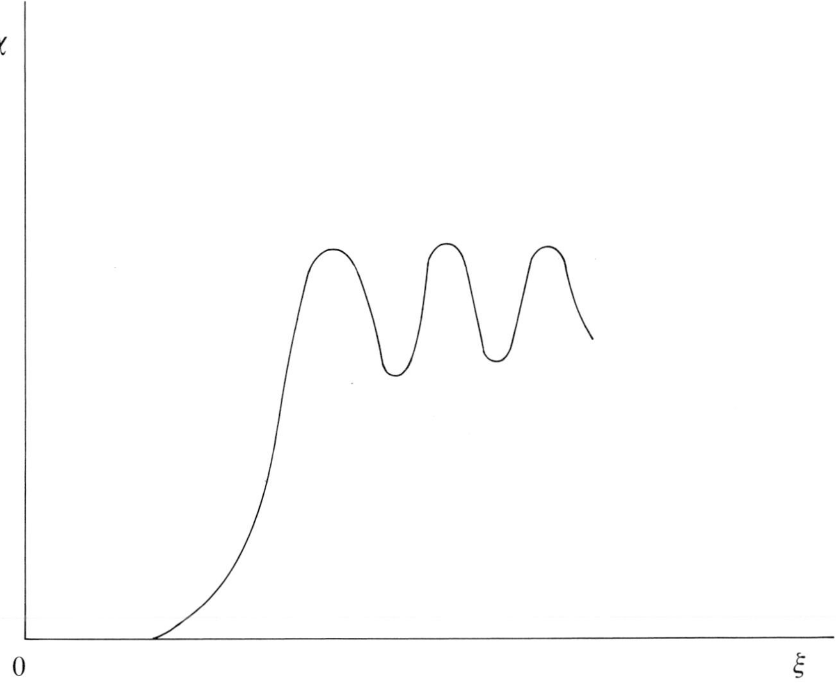

Figure 3.2. The potential variation in the wave.

(ii) Solitary Waves

The ion motions are governed by the following equations:

$$\frac{\partial n}{\partial t} + \frac{\partial}{\partial x}(nv) = 0 \tag{10}$$

$$\frac{\partial V}{\partial t} + V\frac{\partial V}{\partial x} = -\frac{\partial \phi}{\partial x} \tag{11}$$

$$\frac{\partial^2 \phi}{\partial x^2} = e^{\phi} - n \tag{12}$$

where we have non-dimensionalized the various quantities as follows:

$$\begin{aligned} n' &= \frac{n_i}{n_0}, \quad n_e' = \frac{n_e}{n_0}, \quad v' = \frac{v}{C_s} \\ x' &= \frac{x}{\lambda_D}, \quad t' = \frac{t}{(\omega_{p_i})^{-1}}, \quad \phi' = \frac{\phi}{KT_e/e} \end{aligned} \tag{13}$$

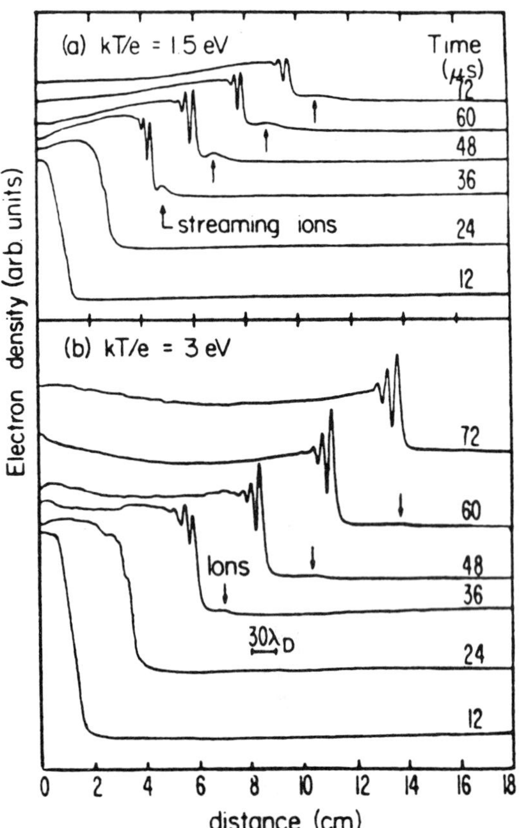

Figure 3.3. Spatial plot of the shock propagation at (a) lower and (b) higher electron temperatures, showing variation in the amount of streaming ions. $T_i = 0.2$ eV; $n_0 = 10^9$ cm^{-3}; initial $\delta n/n_0 = 25\%$; excitation (a) 1 V, (b) 2 V. (Due to Taylor *et al.* [11], by courtesy of The American Physical Society).

and dropped the primes. Here,

$$\omega_{p_i}^2 = \frac{4\pi n_0 e^2}{m_i}.$$

Let us look for solutions for equations (10)—(12), for which

$$x \Rightarrow \infty: \qquad n \Rightarrow 1, \quad \phi \Rightarrow 0, \quad v \Rightarrow 0. \tag{14}$$

Note that if charge neutrality is assumed, equation (11) would become

$$\frac{\partial v}{\partial t} + v\frac{\partial v}{\partial x} = -\frac{1}{n}\frac{\partial n}{\partial x}. \tag{15}$$

Equations (10) and (15) lead to wave solutions with unlimited steepening of an initial perturbation until breaking occurs. However, when spatial gradients become large it is no longer permissible to neglect $\partial^2\phi/\partial x^2$ in equation (12), and it is this dispersive effect of deviations from charge neutrality which eventually limits the build-up of short-wavelength components of the disturbance.

Let us look for travelling-wave solutions with dependence on x and t through the combination $\xi = x - Mt$, (note that $M = V_0/C_s$, V_0 being the speed of propagation of the wave). Then, equations (10) and (11) give

$$n = \frac{M}{M - v} \tag{16}$$

$$(M - v)^2 = M^2 - 2\phi. \tag{17}$$

Using (16) and (17), equation (12) gives

$$\frac{1}{2}\left(\frac{\mathrm{d}\phi}{\mathrm{d}\xi}\right)^2 = \left[-(M^2 + 1) + e^{\phi} + M^2\left(1 - \frac{2\phi}{M^2}\right)^{1/2}\right] \tag{18}$$

where we have used the fact that the quasipotential $V(\phi)$ satisfies

$$\phi = 0: \quad V(\phi) = 0.$$

Note that the quasipotential $V(\phi)$ given by (18) is the same as in (8). If $1 < M^2 < 1.6$, $V(\phi)$ has a shape shown in Figure 3.1. This shape gives rise to the existence of a solitary wave, as mentioned previously. In order to get an analytic expression for it, let us consider

$$M = 1 + \Delta, \quad \Delta \ll 1 \tag{19}$$

then (18) reduces to

$$\left(\frac{\mathrm{d}\phi}{\mathrm{d}\xi}\right)^2 = \frac{2}{3}\,\phi^2(3\Delta - \phi) \tag{20}$$

from which,

$$\phi = 3\Delta\,\mathrm{sech}^2\left[\sqrt{\frac{\Delta}{2}}\,(x - Mt)\right] \tag{21}$$

which represents a small-amplitude solitary wave (Figure 3.4), travelling slightly faster than the ion-acoustic speed C_s.

Solution (21), obtained in the weakly-nonlinear regime ($\phi_{\max} \ll 1$) belongs to an important class of exact solutions of certain nonlinear partial differential equations. In order to derive such a partial differential equation, note that the argument of the sech function in (21) can be rewritten as

$$\sqrt{\tfrac{1}{2}}\,[\sqrt{\Delta}\,(x - t) - \Delta^{3/2}t]$$

indicating two appropriate scalings in space and time variables (in a frame

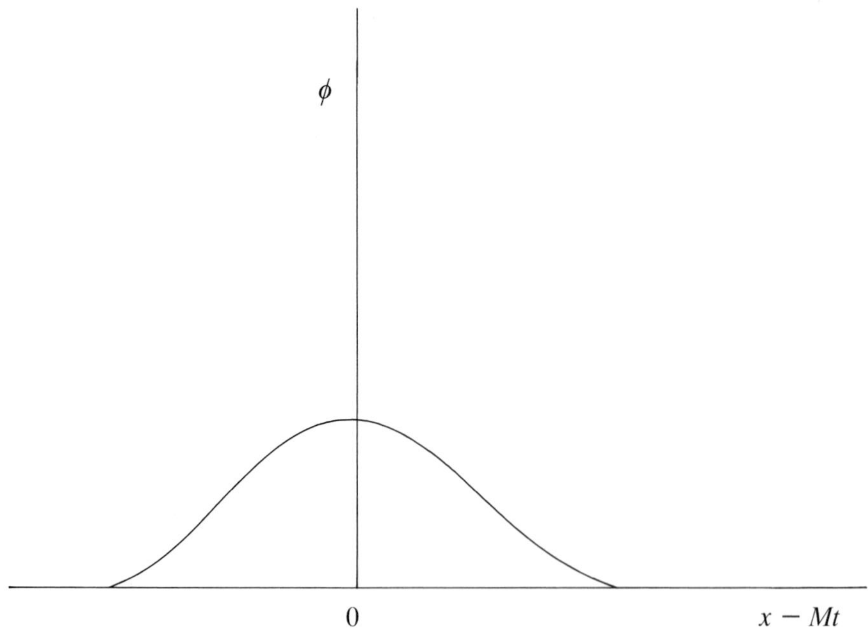

Figure 3.4. A solitary wave.

moving with $M = 1$) required to obtain the equation. This recognition is due to Washimi and Taniuti [4]. Accordingly, let us introduce the following variables

$$\xi = \sqrt{\varepsilon}\,(x - t), \quad \tau = \varepsilon^{3/2} t, \quad \varepsilon \ll 1 \tag{22}$$

so that equations (10)—(12) become:

$$\varepsilon \frac{\partial n}{\partial \tau} - \frac{\partial n}{\partial \xi} + \frac{\partial}{\partial \xi}(nv) = 0 \tag{23}$$

$$\varepsilon \frac{\partial v}{\partial \tau} - \frac{\partial v}{\partial \xi} + v \frac{\partial v}{\partial \xi} = - \frac{\partial \phi}{\partial \xi} \tag{24}$$

$$\varepsilon \frac{\partial^2 \phi}{\partial \xi^2} = e^{\phi} - n. \tag{25}$$

Let us next seek solutions to equations (23)—(25) of the form:

$$\begin{aligned} n &= 1 + \varepsilon n_1 + \varepsilon^2 n_2 + 0(\varepsilon^3) \\ v &= \varepsilon v_1 + \varepsilon^2 v_2 + 0(\varepsilon^3) \\ \phi &= \varepsilon \phi_1 + \varepsilon^2 \phi_2 + 0(\varepsilon^3). \end{aligned} \tag{26}$$

One then obtains from equations (23)—(25):

$0(\varepsilon)$:

$$-\frac{\partial n_1}{\partial \xi} + \frac{\partial v_1}{\partial \xi} = 0 \tag{27}$$

$$-\frac{\partial v_1}{\partial \xi} = -\frac{\partial \phi_1}{\partial \xi} \tag{28}$$

$$\phi_1 = n_1 \tag{29}$$

$0(\varepsilon^2)$:

$$\frac{\partial n_1}{\partial \tau} - \frac{\partial n_2}{\partial \xi} + \frac{\partial v_2}{\partial \xi} + \frac{\partial}{\partial \xi}(n_1 v_1) = 0 \tag{30}$$

$$\frac{\partial v_1}{\partial \tau} - \frac{\partial v_2}{\partial \xi} + v_1 \frac{\partial v_1}{\partial \xi} = -\frac{\partial \phi_2}{\partial \xi} \tag{31}$$

$$\frac{\partial^2 \phi_1}{\partial \xi^2} = \phi_2 + \frac{\phi_1^2}{2} - n_2. \tag{32}$$

One obtains from equations (27)—(29):

$$\phi_1 = v_1 = n_1. \tag{33}$$

Using (33), one derives from equations (30)—(32),

$$\frac{\partial \phi_1}{\partial \tau} + \phi_1 \frac{\partial \phi_1}{\partial \xi} = -\frac{1}{2}\frac{\partial^3 \phi_1}{\partial \xi^3} \tag{34}$$

which is the Korteweg—de Vries equation.

[**Note:** The linear dispersion relation for the ion-acoustic waves is given by Krall and Trivelpiece [119]:

$$\frac{\omega}{k} = C_s(1 + k^2\lambda_D^2)^{-1/2}.$$

If $k\lambda_D \ll 1$, this gives

$$\omega \approx kC_s(1 - \tfrac{1}{2}k^2\lambda_D^2 + \cdots).$$

Observe that the coefficient of the term on the right hand side of equation (34) is the same as the coefficient of k^3 in the linear dispersion relation.]

In order to find a solution to equation (34), let us write it in the form (Scott *et al.* [120])

$$\phi_\tau + \alpha\phi\phi_\xi + \phi_{\xi\xi\xi} = 0. \tag{35}$$

Let us look for steady, progressing wave solutions of the form

$$\phi(\xi, \tau) = \phi(\eta), \quad \eta \equiv \xi - u\tau \tag{36}$$

so that one obtains from equation (35),

$$\phi_\eta(\alpha\phi - u) + \phi_{\eta\eta\eta} = 0. \tag{37}$$

Let us impose the boundary conditions:

$$|\eta| \Rightarrow \infty: \quad \phi, \phi_\eta, \phi_{\eta\eta} \Rightarrow 0. \tag{38}$$

Upon integrating equation (37) and using (38), one obtains

$$\phi_{\eta\eta} = u\phi - \frac{\alpha}{2}\phi^2 \tag{39}$$

and again,

$$\frac{1}{2}\phi_\eta^2 = \frac{u}{2}\phi^2 - \frac{\alpha}{6}\phi^3 \tag{40}$$

from which,

$$\int_{\phi_{\max}}^{\phi} \mathrm{d}\phi \left(\frac{u}{2}\phi^2 - \frac{\alpha}{6}\phi^3\right)^{-1/2} = \sqrt{2}\,\eta. \tag{41}$$

The integral on the left is of the form

$$I = \int \frac{\mathrm{d}\phi}{\phi\sqrt{1 - \beta\phi}}. \tag{42}$$

Put,

$$\psi = \sqrt{1 - \beta\phi}, \quad \beta \equiv \frac{\alpha}{3u} \tag{43}$$

then,

$$I = -2\int \frac{\mathrm{d}\psi}{1 - \psi^2} = \ln\frac{1 - \psi}{1 + \psi}. \tag{44}$$

Thus,

$$\frac{1 - \psi}{1 + \psi} = e^{\sqrt{u}\,\eta} \tag{45}$$

from which,

$$\psi = \frac{1 - e^{\sqrt{u}\,\eta}}{1 + e^{\sqrt{u}\,\eta}}. \tag{46}$$

But, from (43),

$$\phi = \frac{1 - \psi^2}{\beta}. \tag{47}$$

Using (46), (47) becomes

$$\begin{aligned}\phi &= \frac{1}{\beta}\,\frac{4e^{\sqrt{u}\,\eta}}{(1+e^{\sqrt{u}\,\eta})^2} = \frac{1}{\beta}\,\operatorname{sech}^2\frac{\sqrt{u}}{2}\,\eta \\ &= \frac{3u}{\alpha}\operatorname{sech}^2\left[\frac{\sqrt{u}}{2}(\xi - u\tau)\right] \end{aligned} \tag{48}$$

which represents a unidirectional solitary wave. Equation (48) shows that:

(a) The solitary wave moves with velocity faster than the ion-acoustic speed by an amount proportional to one-third of its amplitude;
(b) The width of the solitary wave is inversely proportional to the square root of its amplitude.

Solitary waves are localized waves propagating without change of shape and velocity. The essential quality of a solitary wave is the balance between nonlinearity, which tends to steepen the wavefront in consequence of the increase of wave speed with amplitude, and dispersion which tends to spread the wavefront. Zabusky and Kruskal [6] numerically found that solitary waves preserve their identity and are stable in processes of mutual collisions. The effect of damping of an ion-acoustic solitary wave due to ion-neutral collisions was considered by Ott and Sudan [121] and Ott [122].

Observations of the formation and propagation of ion-acoustic solitons were made by Ikezi *et al.* [8, 9]. The latter observed that in a high electron-to-ion temperature ratio plasma an ion-acoustic pulse broke up into a soliton and a trailing wavetrain. The amplitude and the width of the soliton was related to its velocity in close agreement with the predictions of equation (48). A precursor was found in front of the soliton when the amplitude of the latter is large. The precursor consisted mainly of the ions that were reflected by the potential of the soliton (if the soliton has a finite positive potential ϕ_0, ions in the velocity range $u - \sqrt{2e\phi_0/M} < v < u$ are reflected by the soliton) and carried away the wave energy and this induced an enhanced damping of the solitons. The interaction of two solitons was investigated in the following two cases:

(a) two different-amplitude solitons propagating in the same direction,
(b) two solitons propagating in opposite directions to each other (such an interaction cannot be described by equation (35), however).

In case (a), since the larger-amplitude soliton propagates faster, it will overtake the smaller one. Figure (3.5a) shows the resulting interaction. The wave profile is shown as a function of distance for different times after the excitation of the first pulse. The interaction is depicted in the wave frame moving with the small pulse initially. As the larger pulse overtakes the small pulse, the amplitude of the front pulse increases as the amplitude of the large pulse decreases. Finally, the amplitude of the front pulse is larger than the pulse in back. In Figure (3.5b) a soliton is excited at each end of the main plasma at the same time. The solitons propagate toward each other and interact near the center of the main

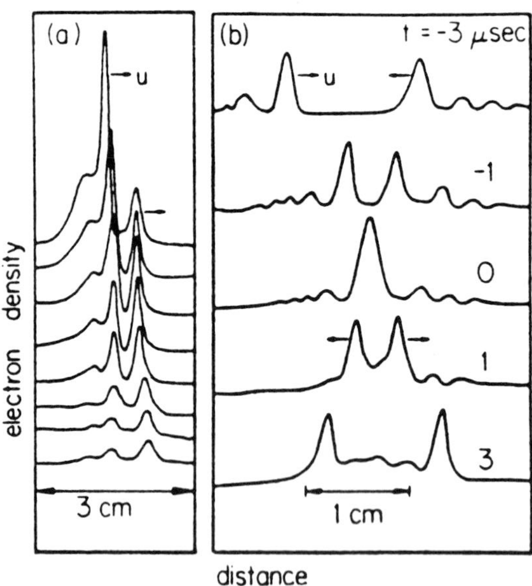

Figure 3.5. Interactions of two solitons. (a) Two solitons propagate in the same direction in the laboratory frame. The figure is depicted in the wave frame such that the smaller soliton is initially stationary. Time differences between each adjacent two curves are 10 μsec. (b) Two solitons propagating in opposite directions to each other, depicted in the laboratory frame. $\lambda_D \approx 2 \times 10^{-2}$ cm. (Due to Ikezi *et al.* [8], by courtesy of The American Physical Society).

plasma. The two solitons add up linearly when they overlap and penetrate through each other.

Further, an initial state of the waveform solution of the Korteweg–de Vries equation is found to show recurrence behavior (Abe *et al.* [123, 124]). Gardner *et al.* [7] have given an ingenious technique to construct an *exact* solution of the initial-value problem for the Korteweg–de Vries equation through a sequence of linear problems. This method of solution involves the following program:

(i) Direct problem: One sets up an appropriate scattering problem in the space variable for the Schrödinger equation (in quantum mechanics!) where the solution $\psi(x, t)$ of the Korteweg–de Vries equation plays the role of the potential; given $\psi(x, 0)$, one finds the scattering data and bound-state eigenfunctions and eigenvalues;
(ii) Time-evolution of the scattering data: This is determined as described by the Korteweg–de Vries equation;
(iii) Inverse problem: Given the scattering data and bound-state eigenfunctions and eigen values, one finds $\psi(x, t)$.

We will give here no account of this inverse scattering method, because several excellent accounts are available in the literature (see for example, Whitham [160]).

(iii) Interacting Solitary Waves

We will make an analytical calculation here of the behavior of two interacting solitary waves. The treatment is due to Whitham [160]. Consider Korteweg–de Vries equation in the form

$$\phi_t + \sigma\phi\phi_x + \phi_{xxx} = 0. \tag{49}$$

Put,

$$\phi = \psi_x. \tag{50}$$

Equation (49) becomes

$$\psi_t + \tfrac{1}{2}\sigma\psi_x^2 + \psi_{xxx} = 0. \tag{51}$$

Next put

$$\sigma\psi = 12(\ln F)_x. \tag{52}$$

Equation (51) becomes

$$F(F_t + F_{xxx})_x - F_x(F_t + F_{xxx}) + 3(F_{xx}^2 - F_x F_{xxx}) = 0. \tag{53}$$

Let us look for a solution to equation (53) of the form

$$F = 1 + \varepsilon F_1 + \varepsilon^2 F_2 + \cdots, \quad \varepsilon \ll 1 \tag{54}$$

so that equation (53) gives

$$(F_{1_t} + F_{1_{xxx}})_x = 0 \tag{55}$$

$$(F_{2_t} + F_{2_{xxx}})_x = -3(F_{1_{xx}}^2 - F_{1_x}F_{1_{xxx}}) \tag{56}$$

$$\begin{aligned}(F_{3_t} + F_{3_{xxx}})_x = &-F_1(F_{2_t} + F_{2_{xxx}})_x + F_{1_x}(F_{2_t} + F_{2_{xxx}}) \\ &- 3(2F_{1_{xx}}F_{2_{xx}} - F_{1_x}F_{2_{xxx}} - F_{2_x}F_{1_{xxx}})\end{aligned} \tag{57}$$

$$\begin{aligned}(F_{4_t} + F_{4_{xxx}})_x = &-F_2(F_{2_{tx}} + F_{2_{xxxx}}) + F_{2_x}(F_{2_t} + F_{2_{xxx}}) \\ &- 3(F_{2_{xx}}^2 - F_{2_x}F_{2_{xxx}}).\end{aligned} \tag{58}$$

One obtains from equation (55), a two-soliton solution

$$\begin{aligned}F_1 &= f_1 + f_2 \\ f_j &= e^{-\alpha_j(x - s_j) + \alpha_j^3 t}\end{aligned} \tag{59}$$

Using (59), equation (56) becomes

$$(F_{2_t} + F_{2_{xxx}})_x = 3\alpha_1\alpha_2(\alpha_2 - \alpha_1)^2 f_1 f_2 \tag{60}$$

from which,

$$F_2 = \frac{(\alpha_2 - \alpha_1)^2}{(\alpha_2 + \alpha_1)^2} f_1 f_2. \tag{61}$$

Using (59) and (61), equation (57) becomes

$$
\begin{aligned}
(F_{3_t} + F_{3_{xxx}})_x = {} & \frac{(\alpha_2 - \alpha_1)^2}{(\alpha_2 + \alpha_1)^2} [-(f_1 + f_2)\{(f_1 f_2)_{tx} + (f_1 f_2)_{xxxx}\} \\
& + (f_{1_x} + f_{2_x})\{(f_1 f_2)_t + (f_1 f_2)_{xxx}\} \\
& - 3\{2(f_{1_{xx}} + f_{2_{xx}})(f_1 f_2)_{xx} - (f_{1_x} + f_{2_x})(f_1 f_2)_{xxx} \\
& - (f_1 f_2)_x (f_{1_{xxx}} + f_{2_{xxx}})\}] \\
= {} & \frac{(\alpha_2 - \alpha_1)^2}{(\alpha_2 + \alpha_1)^2} [-(f_1 + f_2)(6 f_{1_{xx}} f_{2_{xx}} + 3 f_{1_x} f_{2_{xxx}} \\
& + 3 f_{1_{xxx}} f_{2_x}) + (f_{1_x} + f_{2_x})(3 f_{1_x} f_{2_x} + 3 f_{1_{xx}} f_{2_x}) \\
& - 3\{2(f_{1_{xx}} + f_{2_{xx}})(f_{1_{xx}} f_2 + 2 f_{1_x} f_{2_x} + f_1 f_{2_{xx}}) \\
& - (f_{1_x} + f_{2_x})(f_{1_{xxx}} f_2 + 3 f_{1_{xx}} f_{2_x} + 3 f_{1_x} f_{2_{xx}} \\
& + f_1 f_{2_{xxx}}) - (f_{1_x} f_2 + f_1 f_{2_x})(f_{1_{xxx}} + f_{2_{xxx}})\}] \\
= {} & \frac{(\alpha_2 - \alpha_1)^2}{(\alpha_2 + \alpha_1)^2} f_1 f_2 [-3\alpha_1 \alpha_2 (\alpha_1 + \alpha_2)(f_1 + f_2) \\
& + 3\alpha_1 \alpha_2 (\alpha_1 f_1 + \alpha_2 f_2) - 3\{2(\alpha_1 + \alpha_2)(\alpha_1^2 f_1 \\
& + \alpha_2^2 f_2) - (\alpha_1 + \alpha_2)^2 (\alpha_1 f_1 + \alpha_2 f_2) \\
& - (\alpha_1^3 f_1 + \alpha_2^3 f_2)\}] = 0
\end{aligned} \tag{62}
$$

so that

$$F_3 = 0. \tag{63}$$

Using (59), (61) and (63), equation (58) becomes

$$(F_{4_t} + F_{4_{xxx}})_x = 0 \tag{64}$$

so,

$$F_4 = 0. \tag{65}$$

Thus,

$$F_n = 0, \quad n > 2. \tag{66}$$

Therefore,

$$F = 1 + \varepsilon(f_1 + f_2) + \varepsilon^2 \frac{(\alpha_2 - \alpha_1)^2}{(\alpha_2 + \alpha_1)^2} f_1 f_2 \tag{67}$$

is an exact solution!

Thus,

$$\frac{\sigma}{12}\phi = \frac{\left[\varepsilon(\alpha_1^2 f_1 + \alpha_2^2 f_2) + 2\varepsilon^2(\alpha_2-\alpha_1)^2 f_1 f_2 + \varepsilon^3\frac{(\alpha_2-\alpha_1)^2}{(\alpha_2+\alpha_1)^2}(\alpha_2^2 f_1^2 f_2 + \alpha_1^2 f_1 f_2^2)\right]}{\left[1+\varepsilon(f_1+f_2)+\varepsilon^2\frac{(\alpha_2-\alpha_1)^2}{(\alpha_2+\alpha_1)^2} f_1 f_2\right]}. \tag{68}$$

Note that a single solitary wave is given by

$$\frac{\sigma}{12}\phi = \varepsilon\frac{\alpha^2 f}{(1+\varepsilon f)^2}. \tag{69}$$

From now on, we will put $\varepsilon = 1$.
Further

$$\phi = \phi_{\max} = \frac{3\alpha^2}{\sigma}$$

occurs on

$$f = 1, \quad x = s + \alpha^2 t.$$

Next, note that

$$\begin{aligned} &f_1 \approx 1,\ f_2 \ll 1: \quad \frac{\sigma}{12}\phi \approx \frac{\alpha_1^2 f_1}{(1+f_1)^2} \\ &f_1 \approx 1,\ f_2 \gg 1: \quad \frac{\sigma}{12}\phi \approx \frac{\alpha_1^2\tilde{f}_1}{(1+\tilde{f}_1)^2},\quad \tilde{f}_1 \equiv \frac{(\alpha_2-\alpha_1)^2}{(\alpha_2+\alpha_1)^2} f_1. \end{aligned} \tag{70}$$

The latter is a solitary wave with s_1 replaced by $\tilde{s}_1$, where

$$\tilde{s}_1 \equiv s_1 - \frac{1}{\alpha_1}\ln\left(\frac{\alpha_2+\alpha_1}{\alpha_2-\alpha_1}\right)^2$$

which signifies a finite displacement of the profile in the x-direction. Similarly, when $f_2 \approx 1$, and f_1 is either large or small, one has the solitary wave α_2 with or without a shift in s_2. Where $f_1 \approx 1$ and $f_2 \approx 1$, one has the interaction region. Where f_1 and f_2 are both small or large, one has $\sigma\phi \approx 0$.

In order to consider the behavior of two interacting solitary waves, let $\alpha_2 > \alpha_1 > 0$; the solitary wave α_2 is bigger and so moves faster than the wave α_1. As $t \to -\infty$, one has

$$\begin{aligned} &f_1 \approx 1,\ f_2 \ll 1: \quad \text{wave } \alpha_1 \text{ on } x = s_1 + \alpha_1^2 t \\ &f_2 \approx 1,\ f_1 \gg 1: \quad \text{wave } \alpha_2 \text{ on } x = s_2 - \frac{1}{\alpha_2}\ln\left(\frac{\alpha_2+\alpha_1}{\alpha_2-\alpha_1}\right)^2 + \alpha_2^2 t \end{aligned} \tag{71}$$

and elsewhere $\sigma\phi \approx 0$.

As $t \to \infty$, one has

$$f_1 \approx 1, \quad f_2 \gg 1: \qquad \text{wave } \alpha_1 \text{ on } x = s_1 - \frac{1}{\alpha_1} \ln\left(\frac{\alpha_2 + \alpha_1}{\alpha_2 - \alpha_1}\right)^2 + \alpha_1^2 t$$

$$f_2 \approx 1, \quad f_1 \ll 1: \qquad \text{wave } \alpha_2 \text{ on } x = s_2 + \alpha_2^2 t, \tag{72}$$

and elsewhere $\sigma\phi \approx 0$.

Thus, the solitary waves emerge unchanged in form with the original parameters α_1 and α_2, the faster wave α_2 now being ahead of the slower wave α_1. The only remnant of the collision process is a forward shift

$$\frac{1}{\alpha_2} \ln\left(\frac{\alpha_2 + \alpha_1}{\alpha_2 - \alpha_1}\right)^2$$

for the wave α_2, and a backward shift

$$\frac{1}{\alpha_1} \ln\left(\frac{\alpha_2 + \alpha_1}{\alpha_2 - \alpha_1}\right)^2$$

for the wave α_1, from where they would have been had there been no interaction.

(iv) Ion-Acoustic Waves in a Magnetized Plasma

The nonlinear development of ion-acoustic waves in a magnetized plasma has been considered by Zakharov and Kuznetsov [125], Laedke and Spatschek [126, 127], Infeld [128], Infeld and Frycz [129] and Shivamoggi [130]. Here the dispersion arising from charge separation as well as finite-gyroradius effects can balance the nonlinearity. We will now derive the equation describing the nonlinear development of ion-acoustic waves in a magnetized plasma under the restrictions of small wave-amplitude and weak dispersion.

Consider a plasma in a strong magnetic field $\mathbf{B} = B\hat{\mathbf{i}}_z$ with cold ions and hot electrons ($T_e \gg T_i$). The ion motions are now governed by the following equations:

$$\frac{\partial n}{\partial t} + \nabla \cdot (n\mathbf{V}) = 0 \tag{73}$$

$$\frac{\partial \mathbf{V}}{\partial t} + (\mathbf{V} \cdot \nabla)\mathbf{V} = -\frac{e}{m_i} \nabla\phi + \mathbf{V} \times \mathbf{\Omega}_i \tag{74}$$

$$\nabla^2 \phi = -4\pi e(n - n_e) \tag{75}$$

$$n_e = n_0 e^{e\phi/KT_e} \tag{76}$$

where,

$$\mathbf{\Omega}_i \equiv \frac{e\mathbf{B}}{m_i c}.$$

Let us non-dimensionalize the various quantities as follows:

$$n' = \frac{n}{n_0}, \quad n_e' = \frac{n_e}{n_0}, \quad \phi' = \frac{\phi}{KT_e/e}$$
$$x' = \frac{x}{\rho}, \quad z' = \frac{z}{\lambda_D}, \quad t' = \frac{t}{\lambda_D/C_s}, \quad \rho = \frac{C_s}{\Omega_i} \tag{77}$$

We drop the primes henceforth. Let us approximate the x-component of the ion-velocity by the polarization drift:

$$V_x = -\frac{\partial^2 \phi}{\partial x \partial t}. \tag{78}$$

Then, equations (73)—(76) become:

$$\frac{\partial n}{\partial t} - \frac{\partial}{\partial x}\left(n \frac{\partial^2 \phi}{\partial x \partial t}\right) + \frac{\partial}{\partial z}(nw) = 0 \tag{79}$$

$$\frac{\partial w}{\partial t} - \frac{\partial^2 \phi}{\partial x \partial t}\frac{\partial w}{\partial x} + w\frac{\partial w}{\partial z} = -\frac{\partial \phi}{\partial z} \tag{80}$$

$$\alpha \frac{\partial^2 \phi}{\partial x^2} + \frac{\partial^2 \phi}{\partial z^2} = e^{\phi} - n \tag{81}$$

where

$$\alpha \equiv \frac{\lambda_D^2}{\rho^2}.$$

Let us now introduce new independent variables:

$$\xi = \varepsilon^{1/2}(z - t), \quad \eta = \varepsilon^{1/2} x, \quad \tau = \varepsilon^{3/2} t \tag{82}$$

where ε is a small parameter characterizing the typical amplitude of the waves. Equations (79)—(81) then become:

$$\varepsilon \frac{\partial n}{\partial \tau} - \frac{\partial n}{\partial \xi} - \varepsilon \frac{\partial}{\partial \eta}\left[n\left(\varepsilon \frac{\partial^2 \phi}{\partial \eta \partial \tau} - \frac{\partial^2 \phi}{\partial \eta \partial \xi}\right)\right]$$
$$+ \frac{\partial}{\partial \xi}(nw) = 0 \tag{83}$$

$$\varepsilon \frac{\partial w}{\partial \tau} - \frac{\partial w}{\partial \xi} - \varepsilon \left(\varepsilon \frac{\partial^2 \phi}{\partial \eta \partial \tau} - \frac{\partial^2 \phi}{\partial \eta \partial \xi} \right) \frac{\partial w}{\partial \eta}$$

$$+ w \frac{\partial w}{\partial \xi} = - \frac{\partial \phi}{\partial \xi} \tag{84}$$

$$\varepsilon \left(\alpha \frac{\partial^2 \phi}{\partial \eta^2} + \frac{\partial^2 \phi}{\partial \xi^2} \right) = e^{\phi} - n. \tag{85}$$

Let us seek solutions of the form:

$$n = 1 + \varepsilon n_1 + \varepsilon^2 n_2 + \cdots$$
$$\phi = \varepsilon \phi_1 + \varepsilon^2 \phi_2 + \cdots$$
$$w = \varepsilon w_1 + \varepsilon^2 w_2 + \cdots . \tag{86}$$

Equations (83)—(85) then give:

$0(\varepsilon)$:

$$- \frac{\partial n_1}{\partial \xi} + \frac{\partial w_1}{\partial \xi} = 0 \tag{87}$$

$$- \frac{\partial w_1}{\partial \xi} = - \frac{\partial \phi_1}{\partial \xi} \tag{88}$$

$$n_1 = \phi_1 \tag{89}$$

$0(\varepsilon^2)$:

$$\frac{\partial n_1}{\partial \tau} - \frac{\partial n_2}{\partial \xi} + \frac{\partial^3 \phi_1}{\partial \eta^2 \partial \xi} + \frac{\partial w_2}{\partial \xi} + \frac{\partial}{\partial \xi} (n_1 w_1) = 0 \tag{90}$$

$$\frac{\partial w_1}{\partial \tau} - \frac{\partial w_2}{\partial \xi} + w_1 \frac{\partial w_1}{\partial \xi} = - \frac{\partial \phi_2}{\partial \xi} \tag{91}$$

$$\alpha \frac{\partial^2 \phi_1}{\partial \eta^2} + \frac{\partial^2 \phi_1}{\partial \xi^2} = \phi_2 + \frac{\phi_1^2}{2} - n_2. \tag{92}$$

We obtain from equations (87)—(89):

$$n_1 = \phi_1 = w_1. \tag{93}$$

Using (93), we may derive the Zakharov—Kuznetsov equation from equations (90)—(92):

$$\frac{\partial \phi_1}{\partial \tau} + \phi_1 \frac{\partial \phi_1}{\partial \xi} + \frac{1}{2} \left[\frac{\partial^3 \phi_1}{\partial \xi^3} + (1 + \alpha) \frac{\partial^3 \phi_1}{\partial \eta^2 \partial \xi} \right] = 0 \tag{94}$$

which is a generalization of the Korteweg—de Vries equation for two-dimensions.

The linear dispersion relation for the ion-acoustic waves in strong magnetic fields (i.e., $\omega \ll \Omega_i$) is given by (Krall and Trivelpiece [119]):

$$\omega = k_z C_s [1 + (k_z^2 + k_x^2)\lambda_D^2 + k_x^2 \rho^2]^{-1/2} \tag{95}$$

which for $k_z\lambda_D$, $k_x\lambda_D$, $k_x\rho \ll 1$ gives:

$$\omega \approx k_z C_s [1 - \tfrac{1}{2} k_z^2 \lambda_D^2 - \tfrac{1}{2}(\lambda_D^2 + \rho^2) k_x^2]. \tag{96}$$

When ω, k_z and k_x are reinterpreted as the operators $i\,\partial/\partial t$, $-i\,\partial/\partial z$ and $-i\,\partial/\partial x$, respectively, (96) reproduces the linear part of equation (94).

It may be noted that unlike the Korteweg—de Vries equation, the Zakharov—Kuznetsov equation (94) does not appear to be an integrable system. The solitary-wave solutions belonging to equation (94) appears to possess only a finite number of polynomial conservation laws (three have been explicitly found so far [128]).

IV

Parametric Excitations of Plasma Waves

Parametric amplifications of plasma waves result due to a periodic modulation of a parameter which characterizes the plasma waves. It may occur in two ways — either due to wave-wave interactions or due to temporal variations in the electromagnetic fields in the presence of a large-amplitude pump wave. We will consider in the following examples of both types of excitations.

(i) Parametric Excitation of Electromagnetic Waves by an Oscillating Electric Field

We will consider here the parametric excitation of transverse electromagnetic waves by a high-frequency electric field (Montgomery and Alexeff [31]). The parametric excitation occurs through a coupling between the longitudinal mode and the transverse mode of the plasma-response field.

Consider a spatially-uniform, cold plasma subjected to a high-frequency pump electric field

$$\mathbf{E}^{(0)} = E_0 \sin \omega_0 t \, \hat{\mathbf{i}}_z$$

the plasma particles then execute oscillations in this pump field with velocities

$$\mathbf{V}_s^{(0)} = -V_s^{(0)} \cos \omega_0 t \, \hat{\mathbf{i}}_z \tag{1}$$

where,

$$V_s^{(0)} = \frac{e_s E_0}{m_s \omega_0} .$$

The subscript s refers to the ions and electrons, and the superscript 0 refers to the equilibrium state.

Note that the equilibrium solution (1) is not consistent with the cold-plasma equations, and one requires the introduction of an external current to make (1) self-consistent.

Consider perturbations linearised about the time-dependent equilibrum state given by (1); they are governed by

$$\frac{\partial n_s^{(1)}}{\partial t} + n_s^{(0)} \nabla \cdot \mathbf{V}_s^{(1)} + \mathbf{V}_s^{(0)} \cdot \nabla n_s^{(1)} = 0 \tag{2}$$

$$\frac{\partial \mathbf{V}_s^{(1)}}{\partial t} + \mathbf{V}_s^{(0)} \cdot \nabla \mathbf{V}_s^{(1)} = \frac{e_s}{m_s}\left(\mathbf{E}^{(1)} + \frac{1}{c}\,\mathbf{V}_s^{(0)} \times \mathbf{B}^{(1)}\right) \tag{3}$$

$$\nabla \cdot \mathbf{E}^{(1)} = 4\pi \sum_s n_s^{(1)} e_s \tag{4}$$

$$\nabla \cdot \mathbf{B}^{(1)} = 0 \tag{5}$$

$$\nabla \times \mathbf{E}^{(1)} = -\frac{1}{c}\,\frac{\partial \mathbf{B}^{(1)}}{\partial t} \tag{6}$$

$$\nabla \times \mathbf{B}^{(1)} = \frac{1}{c}\,\frac{\partial \mathbf{E}^{(1)}}{\partial t} + \frac{4\pi}{c} \sum_s e_s [n_s^{(0)} \mathbf{V}_s^{(1)} + n_s^{(1)} \mathbf{V}_s^{(0)}] \tag{7}$$

where the superscript 1 refers to the perturbations. Consider waves propagating perpendicular to the pump field

$$q^{(1)} \sim e^{i\mathbf{k} \cdot \mathbf{x}}, \quad \mathbf{k} \cdot \mathbf{E}^{(0)} = 0. \tag{8}$$

Using (8), and assuming that the pump field $\mathbf{E}^{(0)}$ is weak, equations (2)—(7) give

$$\frac{\partial n_s^{(1)}}{\partial t} + n_s^{(0)}\, i\mathbf{k} \cdot \mathbf{V}_s^{(1)} = 0 \tag{9}$$

$$\frac{\partial \mathbf{V}_s^{(1)}}{\partial t} = \frac{e_s}{m_s}\left(\mathbf{E}^{(1)} + \frac{\varepsilon}{c}\,\mathbf{V}_s^{(0)} \times \mathbf{B}^{(1)}\right) \tag{10}$$

$$i\mathbf{k} \cdot \mathbf{E}^{(1)} = 4\pi \sum_s e_s n_s^{(1)} \tag{11}$$

$$i\mathbf{k} \cdot \mathbf{B}^{(1)} = 0 \tag{12}$$

$$i\mathbf{k} \times \mathbf{E}^{(1)} = -\frac{1}{c}\,\frac{\partial \mathbf{B}^{(1)}}{\partial t} \tag{13}$$

$$i\mathbf{k} \times \mathbf{B}^{(1)} = \frac{1}{c}\,\frac{\partial \mathbf{E}^{(1)}}{\partial t} + \frac{4\pi}{c} \sum_s e_s [n_s^{(0)} \mathbf{V}_s^{(1)} + \varepsilon n_s^{(1)} \mathbf{V}_s^{(0)}] \tag{14}$$

where ε is a small parameter characterizing the weak pump field.

Let,

$$\begin{aligned}
\mathbf{k} &= k\hat{\mathbf{i}}_x, \quad n_s^{(1)} = n_{s_L}^{(1)} \\
\mathbf{E}^{(1)} &= E_L^{(1)}\hat{\mathbf{i}}_x + E_T^{(1)}\hat{\mathbf{i}}_y \\
\mathbf{B}^{(1)} &= B_T^{(1)}\hat{\mathbf{i}}_z \\
\mathbf{V}_s^{(1)} &= V_{s_L}\hat{\mathbf{i}}_x + V_{s_T}\hat{\mathbf{i}}_y
\end{aligned} \tag{15}$$

so that equations (9)—(14) give:

$$\frac{\partial n_{s_L}^{(1)}}{\partial t} + n_s^{(0)} ikV_{s_L}^{(1)} = 0 \tag{16}$$

$$\frac{\partial V_{s_L}^{(1)}}{\partial t} = \frac{e_s}{m_s} E_L^{(1)} \tag{17}$$

$$\frac{\partial V_{s_T}^{(1)}}{\partial t} = \frac{e_s}{m_s} E_T^{(1)} \tag{18}$$

$$ikE_L^{(1)} = 4\pi \sum_s e_s n_{s_L}^{(1)} \tag{19}$$

$$ikE_T^{(1)} = -\frac{1}{c}\frac{\partial B_T^{(1)}}{\partial t} \tag{20}$$

$$-ikB_T^{(1)} = \frac{1}{c}\frac{\partial E_T^{(1)}}{\partial t} + \frac{4\pi}{c}\sum_s e_s n_s^{(0)} V_{s_T}^{(1)}. \tag{21}$$

One derives from equations (16), (17) and (19),

$$\left(\frac{\partial^2}{\partial t^2} + \omega_p^2\right) E_L^{(1)} = -\varepsilon \sum_s \frac{\omega_{p_s}^2}{c} V_s^{(0)} B_T^{(1)} \cos \omega_0 t \tag{22}$$

and from equations (18), (20) and (21),

$$\left(\frac{\partial^2}{\partial t^2} + \Omega_k^2\right) E_T^{(1)} = \varepsilon \sum_s 4\pi e_s V_s^{(0)} \frac{\partial}{\partial t} (n_{s_L}^{(1)} \cos \omega_0 t) \tag{23}$$

where,

$$\Omega_k^2 = \omega_p^2 + k^2c^2, \quad \omega_p^2 = \sum_s \omega_{p_s}^2, \quad \omega_{p_s}^2 = \frac{4\pi n_s^{(0)} e_s^2}{m_s}.$$

In equations (22) and (23), observe the coupling due to the pump field of the longitudinal and the transverse waves.

The problem in question is characterized by the disparate time scales:

(a) The fast-time scale corresponding to the period of the excited oscillations;
(b) The slow-time scale characterizing the rate at which energy is fed into the excited oscillations; this depends on the strength of the pump field.

Therefore, we use the method of multiple time scales to treat equations (16)–(21). Accordingly, let us seek a solution to equations (16)–(21), of the form

$$q^{(1)}(t;\varepsilon) = q^{(1)}_{(0)}(t,\tilde{t}) + \varepsilon q^{(1)}_{(1)}(t,\tilde{t}) + \cdots \tag{24}$$

where $\tilde{t} \equiv \varepsilon t$ represents the slow-time scale.

Using (24), equations (16)–(21) give:

$$\begin{aligned}
E^{(1)}_{L_{(0)}} &= A^+_L e^{i\omega_p t} + A^-_L e^{-i\omega_p t}\\
E^{(1)}_{T_{(0)}} &= A^+_T e^{i\Omega_k t} + A^-_T e^{-i\Omega_k t}\\
B^{(1)}_{T_{(0)}} &= -\frac{kc}{\Omega_k}\,(A^+_T e^{i\Omega_k t} - A^-_T e^{-i\Omega_k t})\\
V^{(1)}_{s_{L_{(0)}}} &= \frac{e_s}{im_s\omega_p}\,(A^+_L e^{i\omega_p t} - A^-_L e^{-i\omega_p t})\\
V^{(1)}_{s_{T_{(0)}}} &= \frac{e_s}{im_s\Omega_k}\,(A^+_T e^{i\Omega_k t} - A^-_T e^{-i\Omega_k t})\\
e_s n^{(1)}_{s_{L_{(0)}}} &= \frac{ik}{4\pi}\,\frac{\omega^2_{p_s}}{\omega^2_p}\,(A^+_L e^{i\omega_p t} + A^-_L e^{-i\omega_p t})
\end{aligned} \tag{25}$$

where $A^\pm_L$ and $A^\pm_T$ are functions of $\tilde{t}$.

Using (24) and (25), equations (22) and (23) give:

$$\begin{aligned}
\left(\frac{\partial^2}{\partial t^2} + \omega^2_p\right) E^{(1)}_{L_{(1)}} &= -2i\omega_p\left(\frac{\partial A^+_L}{\partial \tilde{t}}\, e^{i\omega_p t} - \frac{\partial A^-_L}{\partial \tilde{t}}\, e^{-i\omega_p t}\right)\\
&\quad - \frac{k}{2\Omega_k}\left(\sum_s \omega^2_{p_s} V^{(0)}_s\right)(A^+_T e^{i\Omega_k t} - A^-_T e^{-i\Omega_k t}) \times\\
&\quad \times (e^{i\omega_0 t} + e^{-i\omega_0 t})
\end{aligned} \tag{26}$$

$$\begin{aligned}
\left(\frac{\partial^2}{\partial t^2} + \Omega^2_k\right) E^{(1)}_{T_{(1)}} &= -2i\Omega_k\left(\frac{\partial A^+_T}{\partial \tilde{t}}\, e^{i\Omega_k t} + \frac{\partial A^-_T}{\partial \tilde{t}}\, e^{-i\Omega_k t}\right)\\
&\quad + ik\sum_s \frac{\omega^2_{p_s}}{2\omega^2_p}\, V^{(0)}_s\, \frac{\partial}{\partial t}\,[(A^+_L e^{i\omega_p t} + A^-_L e^{-i\omega_p t}) \times\\
&\quad \times (e^{i\omega_0 t} + e^{-i\omega_0 t})].
\end{aligned} \tag{27}$$

Equations (26) and (27) develop secular terms on the right hand side if $\omega_0 - \omega_p \approx \Omega_k$. The removal of these secular terms requires

$$\pm 2i\omega_p \frac{\partial A_L^\pm}{\partial \tilde{t}} = \pm \frac{k}{2\Omega_k} \left(\sum_s \omega_{p_s}^2 V_s^{(0)} \right) A_T^\mp e^{\pm i\Delta_k \tilde{t}} \tag{28}$$

$$\pm 2i\Omega_k \frac{\partial A_T^\pm}{\partial \tilde{t}} = \mp k(\omega_0 - \omega_p) \left(\sum_s \frac{\omega_{p_s}^2 V_s^{(0)}}{2\omega_p^2} \right) A_L^\mp e^{\pm i\Delta_k \tilde{t}}, \tag{29}$$

where,

$$\varepsilon \Delta_k \equiv \omega_0 - \omega_p - \Omega_k \tag{30}$$

represents the frequency-mismatch. Observe that the longitudinal and the transverse modes propagating in the opposite directions are coupled parametrically.

From equations (28) and (29), one derives

$$\frac{\partial^2 A_T^\pm}{\partial \tilde{t}^2} \mp i\Delta_k \frac{\partial A_T^\pm}{\partial \tilde{t}} - \frac{k^2}{16\Omega_k \omega_p^3} \left(\sum_s \omega_{p_s}^2 V_s^{(0)} \right)^2 A_T^\pm = 0. \tag{31}$$

Equation (31) gives for an exponentially-growing solution (or for parametric excitation),

$$\frac{k^2}{4\Omega_k \omega_p^3} \left(\sum_s \omega_{p_s}^2 V_s^{(0)} \right)^2 > \Delta_k^2 \tag{32}$$

for a given mismatch in frequency Δ_k, (32) gives the threshold value of the intensity of the pump field $\mathbf{E}^{(0)}$ to cause parametric excitation of transverse electromagnetic waves.

(ii) Wave-Wave Interactions

When two waves with frequencies ω_1 and wave vectors $\mathbf{k}_1$ and $\mathbf{k}_2$, respectively, are present, the nonlinear terms in the plasma equations may contain the product of the wave-amplitudes, *viz.*,

$$\exp[i(\mathbf{k}_1 - \mathbf{k}_2) \cdot \mathbf{x} - i(\omega_1 - \omega_2)t]$$

which is a beat-frequency wave. If the frequency $\omega_1 - \omega_2 = \omega_3$ and the wavevector $\mathbf{k}_1 - \mathbf{k}_2 = \mathbf{k}_3$ of the beat happen to be a normal mode of the plasma, i.e.,

$$\varepsilon(\omega_3, \mathbf{k}_3) = 0$$

$\varepsilon(\omega, \mathbf{k})$ being the lowest-order dielectric constant of the plasma, then this mode will be generated by the interaction of the first two waves. An exchange of energy and momentum takes place among these three waves and the process is called resonant wave interaction. If the wave amplitudes are small, the interactions are selective and weak: selective because only certain combinations of wave components are capable of a significant energy exchange and weak because even for these, the interaction time is large compared with a typical wave period. The smaller the wave amplitudes, the larger will be the interaction time. If the wave $(\omega_1, \mathbf{k}_1)$ corresponds to a driving pump field, then the waves $(\omega_2, \mathbf{k}_2)$, $(\omega_3, \mathbf{k}_3)$ will grow at the expense of the pump $(\omega_1, \mathbf{k}_1)$.

Some of the first experiments on resonant mode-mode coupling were done by Phelps *et al.* [29], Chang and Porkolab [28], and Franklin *et al.* [30]. In these experiments, the frequency and wave-vector selection rules were verified. Chang and Porkolab [28] studied a backscatter type of decay of electron Bernstein waves, and Figure 4.1 shows the decay spectrum and typical experimentally observed decay frequencies and wave numbers plotted in the theoretical linear dispersion diagram. From Figure 4.1 we see that both the frequency and the wave-vector selection rules for resonant mode-mode coupling are satisfied. Phelps *et al.* [29] performed experiments demonstrating resonant nonlinear electromagnetic wave excitation of electron-plasma and ion-acoustic waves in a plasma column and verified the wavenumber selection rules. (The wave coupling took place in the volume of the column and not in the boundaries or in a locally resonant layer.) Franklin *et al.* [30] considered the decay of an electron plasma wave into an ion-acoustic wave and another electron-plasma wave.

A method of determining satisfaction of the resonance conditions

$$\omega_1 = \omega_2 + \omega_3, \quad \mathbf{k}_1 = \mathbf{k}_2 + \mathbf{k}_3$$

was introduced by Peierls [131] in connection with phonon-phonon interaction in solids. In this method, the portion of the dispersion diagram (Figure 4.2) corresponding to (ω_3, k_3) is drawn with origin (ω_2, k_2). If the terminal point of this displaced curve lies on the original dispersion curve then there exists a resonant triad (Figure 4.3).

It is easy to see using this method that resonant triads of electron waves or of electromagnetic waves do not exist. On the other hand, it is possible to find resonant triads involving two electron waves and an ion wave. Figures 4.4—4.6 show some examples of permissible resonant triads. In Figure 4.4, a large-amplitude electron wave $(\omega_0, \mathbf{k}_0)$ decays into a backward moving electron wave $(\omega_2, \mathbf{k}_2)$ and an ion wave $(\omega_1, \mathbf{k}_1)$. In Figure 4.5, an electromagnetic wave $(\omega_0, \mathbf{k}_0)$ decays into an electron wave $(\omega_2, \mathbf{k}_2)$ and an ion wave $(\omega_1, \mathbf{k}_1)$ moving in opposite directions. In Figure 4.6, an electromagnetic wave $(\omega_0, \mathbf{k}_0)$ excites another electromagnetic wave $(\omega_2, \mathbf{k}_2)$ moving in the opposite direction and an ion wave $(\omega_1, \mathbf{k}_1)$.

(*a*) *Nonlinear Resonant Interactions Between Two Electromagnetic Waves and a Langmuir Wave*

The nonlinear resonant interactions between two transverse electromagnetic waves and a longitudinal electron plasma wave in a warm electron-plasma have

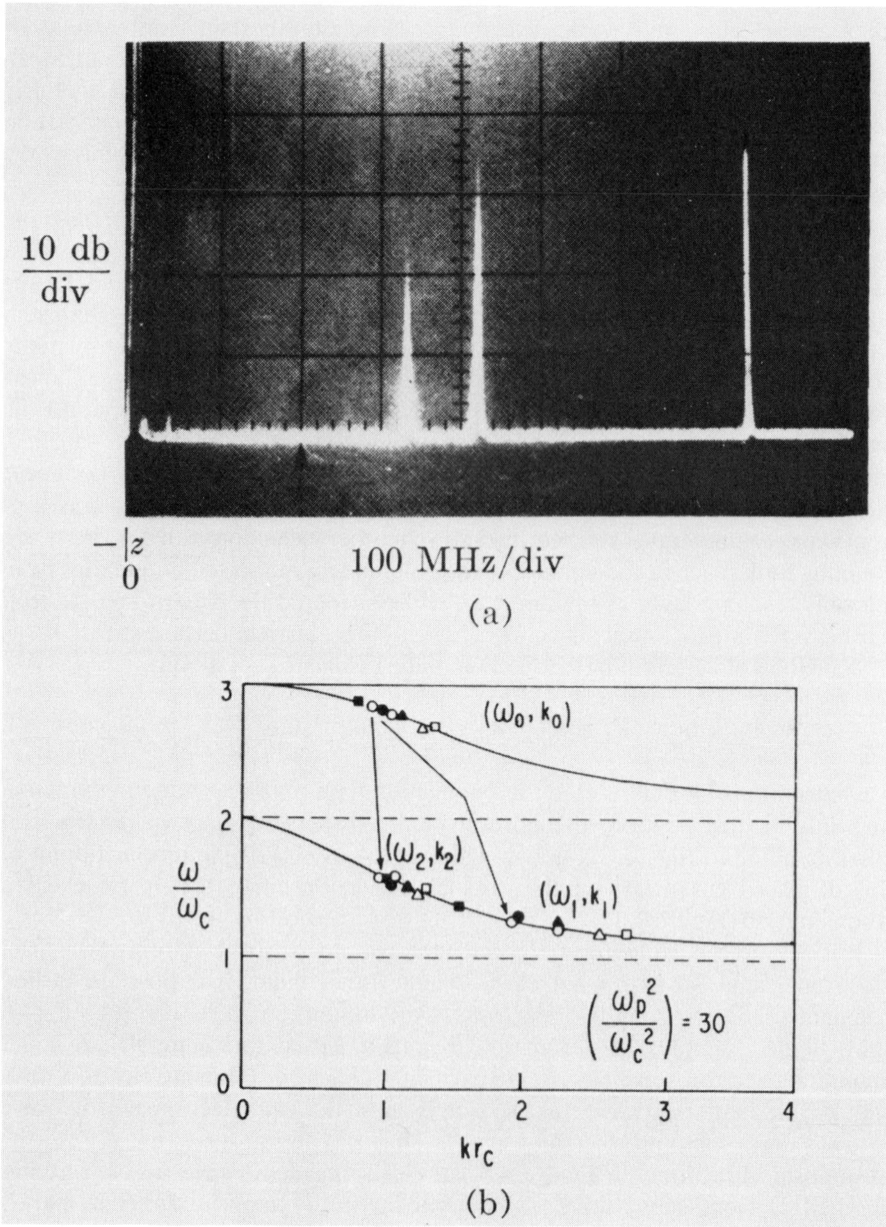

Figure 4.1. (a) Typical decay spectrum; f_1 = 335 MHz, f_2 = 420 MHz, f_0 = 758 MHz. The pump signal (f_0) is greatly attenuated by a filter. ω_c (cyclotron frequency). (b) Dispersion relation and decay modes. Each set of three identical symbols represents a pair of decay modes like that indicated by the arrows. There are seven pairs of decay modes shown in the figure. (Due to Chang and Porkolab [28], by courtesy of The American Institute of Physics.)

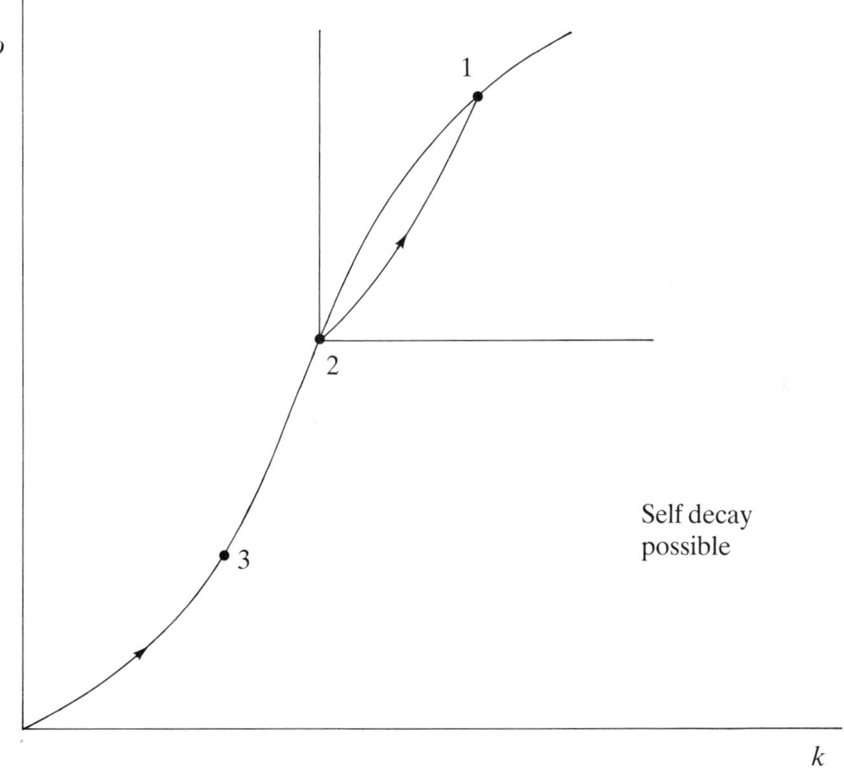

Figure 4.2. Peierls' construction.

been discussed by Sjolund and Stenflo [14], Weiland and Wilhelmsson [132] and Shivamoggi [15].

Consider a plasma comprised of warm electrons and immobile ions constituting a uniform background of positive charge (because we are considering high-frequency waves). The unperturbed plasma is homogeneous in space and there are no external fields acting on the plasma. The equations governing wave motions in such a plasma are:

$$\frac{\partial n}{\partial t} + N_0 \nabla \cdot \mathbf{V} = -\nabla \cdot (n\mathbf{V}) \tag{33}$$

$$\frac{\partial \mathbf{V}}{\partial t} + \frac{e\mathbf{E}}{m} + \frac{V_T^2}{N_0} \nabla n = -(\mathbf{V} \cdot \nabla)\mathbf{V} - \frac{e}{mc} \mathbf{V} \times \mathbf{B} + \frac{V_{T_e}^2}{N_0} n \nabla n \tag{34}$$

$$\frac{1}{c} \frac{\partial \mathbf{B}}{\partial t} + \nabla \times \mathbf{E} = 0 \tag{35}$$

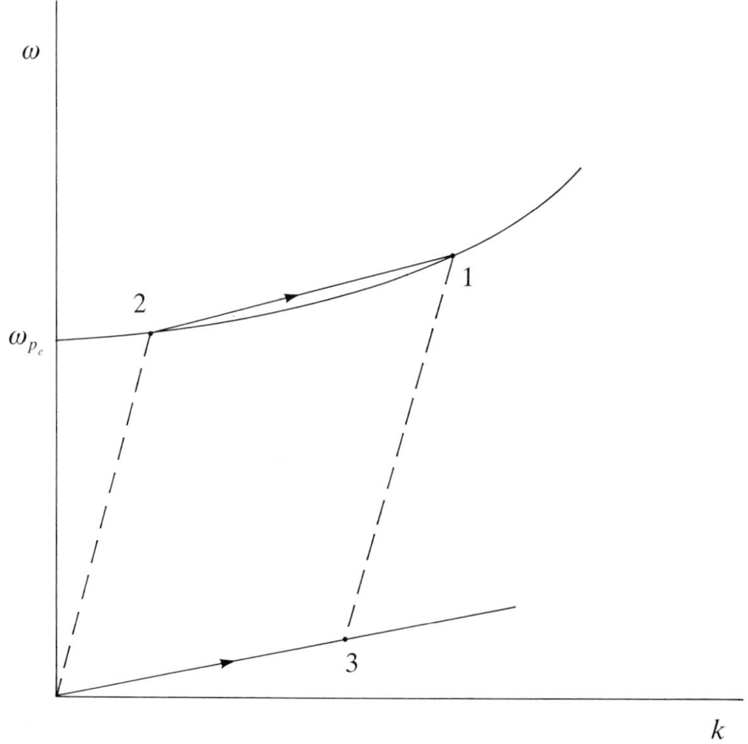

Figure 4.3. Resonant triad $\omega_1 = \omega_2 + \omega_3$.

$$\frac{\partial \mathbf{E}}{\partial t} - c\nabla \times \mathbf{B} - eN_0\mathbf{V} = en\mathbf{V} \tag{36}$$

where N_0 is the number density of electrons (or ions) in the unperturbed state, and V_{T_e} is the thermal speed of electrons. Note that whereas the left hand sides in equations (33)—(36) represent the linearized problem, the right hand sides represent the nonlinear effects.

Let us now consider waves propagating in the x-direction. One may separate equations (33)—(36) into those governing the transverse modes:

$$\frac{\partial v_y}{\partial t} + \frac{eE_y}{m} = \frac{e}{mc} v_x B_z - v_x \frac{\partial v_y}{\partial x} \tag{37}$$

$$\frac{1}{c} \frac{\partial B_z}{\partial t} + \frac{\partial E_y}{\partial x} = 0 \tag{38}$$

$$\frac{\partial E_y}{\partial t} + c \frac{\partial B_z}{\partial x} - eN_0 v_y = env_y \tag{39}$$

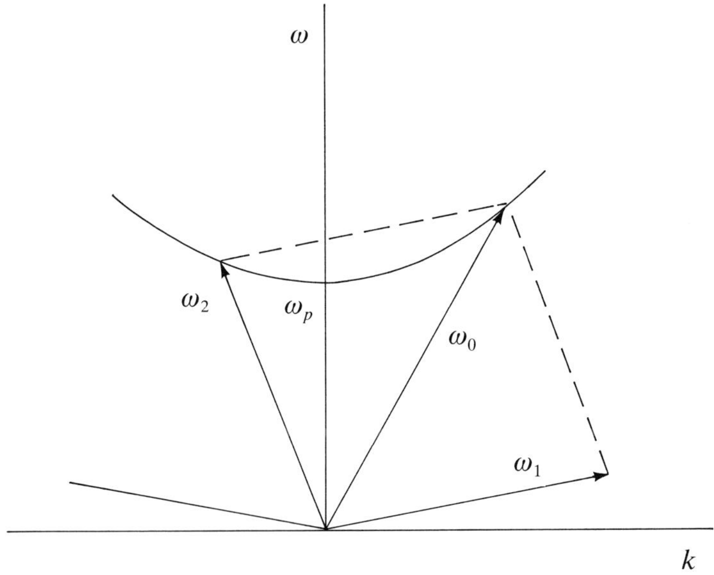

Figure 4.4. A resonant triad. (Due to Chen [141], by courtesy of Plenum Press.)

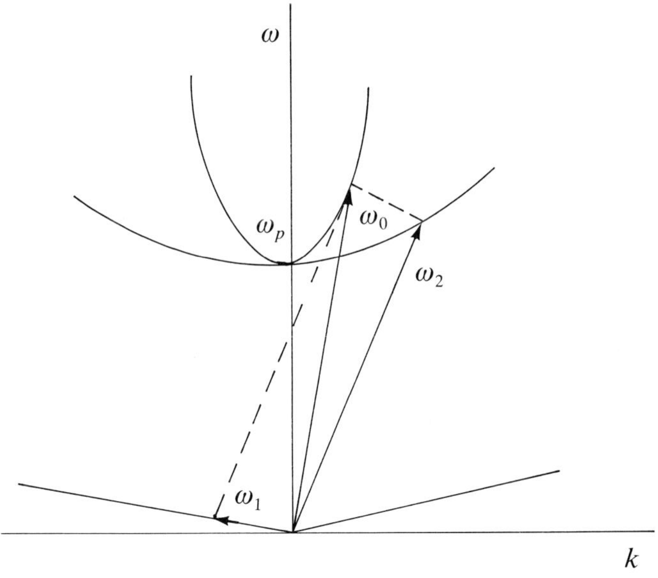

Figure 4.5. A resonant triad. (Due to Chen [141], by courtesy of Plenum Press.)

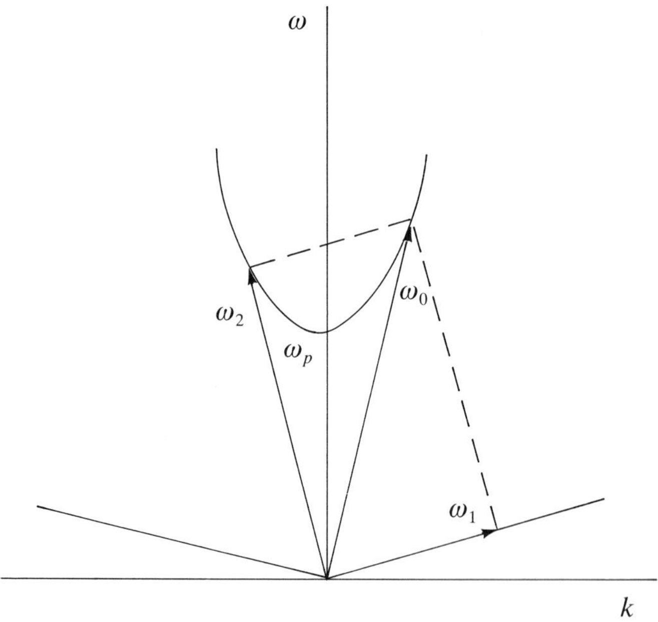

Figure 4.6. A resonant triad. (Due to Chen [141], by courtesy of Plenum Press.)

and the equations governing the longitudinal modes:

$$\frac{\partial n}{\partial t} + N_0 \frac{\partial v_x}{\partial x} = -\frac{\partial}{\partial x}(nv_x) \tag{40}$$

$$\frac{\partial v_x}{\partial t} + \frac{eE_x}{m} + \frac{V_T^2}{N_0}\frac{\partial n}{\partial x} = -\frac{e}{mc} v_y B_z - v_x \frac{\partial v_x}{\partial x} + \frac{V_{T_e}^2}{N_0} n \frac{\partial n}{\partial x} \tag{41}$$

$$\frac{\partial E_x}{\partial t} - eN_0 v_x = env_x \tag{42}$$

where we have assumed that the polarization of the transverse modes is such that

$$E_z = B_y = v_z = 0.$$

We will use the method of coupled modes (Sjolund and Stenflo [14]) to treat this problem. Let us consider linearized transverse modes of the form $\exp(-ik_T x)$ and introduce

$$a_T = v_y + \frac{i\omega_T}{eN_0} E_y + \frac{ik_T c}{eN_0} B_z \tag{43}$$

where,

$$\omega_T^2 \equiv \omega_p^2 + k_T^2 c^2$$

$$\omega_L^2 \equiv \omega_p^2 + k_L^2 V_{T_e}^2, \quad \omega_p^2 = \frac{N_0 e^2}{m}$$

k_L being the wave number of the Langmuir wave.

Using equations (37)—(39), one obtains

$$\frac{\partial a_T}{\partial t} = i\omega_T a_T \tag{44}$$

so that a_T is a normal mode of the linearized problem associated with equations (37)—(39).

Similarly, let us consider linearized longitudinal modes of the form $\exp(-ik_L x)$, and introduce

$$a_L = n + \frac{N_0 \omega_L}{k_L V_{T_e}^2} v_x + \frac{ieN_0}{mk_L V_{T_e}^2} E_x. \tag{45}$$

Using equations (40)—(42), one then obtains,

$$\frac{\partial a_L}{\partial t} = i\omega_L a_L \tag{46}$$

so that a_L is a normal mode of the linearized problem associated with equations (40)—(42).

Note that the transverse and the longitudinal modes are uncoupled in the linearized problem. The coupling between the two modes becomes effective in the nonlinear problem. When the nonlinear terms are included, one obtains from equations (37)—(42),

$$\frac{\partial a_T}{\partial t} - i\omega_T a_T = \frac{e}{mc} v_x B_z - v_x \frac{\partial v_y}{\partial x} + \frac{i\omega_T}{N_0} n v_y \tag{47}$$

$$\frac{\partial a_L}{\partial t} - i\omega_L a_L = -\frac{\partial}{\partial x}(nv_x) - \frac{N_0\omega_L}{k_L V_T^2}\left(v_x \frac{\partial v_x}{\partial x} + \frac{e}{mc} v_y B_z - \frac{V_{T_e}^2}{N_0} n \frac{\partial n}{\partial x}\right) + \frac{ie^2 N_0}{mk_L^2 V_{T_e}^2} n v_x. \tag{48}$$

Let us consider two transverse electromagnetic waves of the form $\exp[-i(k_{T_0} x - \omega_{T_0} t)]$ and $\exp[-i(k_{T_1} x - \omega_{T_1} t)]$ propagation in the x-direction, with

$$\begin{aligned} \omega_{T_0}^2 &= \omega_p^2 + k_{T_0}^2 c^2 \\ \omega_{T_1}^2 &= \omega_p^2 + k_{T_1}^2 c^2. \end{aligned} \tag{49}$$

Due to nonlinear resonant interaction between these two waves, let a Langmuir wave of the form $\exp[-i(k_L x - \omega_L t)]$, propagating in the x-direction, be excited such that

$$\omega_{T_0} - \omega_{T_1} = \omega_L, \quad k_{T_0} - k_{T_1} = k_L. \tag{50}$$

Now, one obtains for the linearized problem.

$$v_x = \frac{k_L V_T^2}{2\omega_L N_0} a_L, \quad n = \frac{k_L^2 V_{T_e}^2}{2\omega_L^2} a_L, \quad E_x = -\frac{iek_L V_{T_e}^2}{2\omega_L^2} a_L$$
$$a_L = a_L(t)e^{i(\omega_L t - k_L x)} + a_L^*(t)e^{-i(\omega_L t - k_L x)} \tag{51}$$

and

$$v_y = \frac{\omega_p^2}{2\omega_T^2} a_T, \quad E_y = -\frac{im}{e}\frac{\omega_p^2}{2\omega_T} a_T$$
$$B_z = -\frac{imk_T c}{e}\frac{\omega_p^2}{2\omega_T^2} a_T \tag{52}$$
$$a_T = a_T(t)e^{i(\omega_T t - k_T x)} + a_T^*(t)e^{-i(\omega_T t - k_T x)}$$

where $a_L(t)$ and $a_T(t)$ are slowly-varying functions of time, and the stars denote complex conjugates.

Using (51) and (52), equations (47) and (48) give, upon keeping only the resonant terms on their right-hand sides:

$$\frac{\partial a_{T_0}}{\partial t} - i\omega_{T_0} a_{T_0} = \frac{ik_L^2 V_{T_e}^2 \omega_p^2 \omega_{T_0}}{4\omega_L^2 \omega_{T_1}^2 N_0} a_{T_1} a_L$$
$$\frac{\partial a_{T_1}}{\partial t} - i\omega_{T_1} a_{T_1} = \frac{ik_L^2 V_{T_e}^2 \omega_p^2 \omega_{T_1}}{4\omega_L^2 \omega_{T_0}^2 N_0} a_{T_0} a_L^* \tag{53}$$
$$\frac{\partial a_L}{\partial t} - i\omega_L a_L = \frac{iN_0 \omega_L \omega_p^4}{4\omega_{T_0}^2 \omega_{T_1}^2 V_{T_e}^2} a_{T_0} a_{T_1}^*.$$

In terms of the electric field components E_y (or E_T) and E_x (or E_L), equation (53) becomes

$$\frac{\partial E_{T_0}}{\partial t} - i\omega_{T_0} E_{T_0} = -\frac{k_L \omega_p^2}{2eN_0 \omega_{T_1}} E_{T_1} E_L$$
$$\frac{\partial E_{T_1}}{\partial t} - i\omega_{T_1} E_{T_1} = \frac{k_L \omega_p^2}{2eN_0 \omega_{T_0}} E_{T_0} E_L^* \tag{54}$$
$$\frac{\partial E_L}{\partial t} - i\omega_L E_L = \frac{k_L \omega_p^4}{2eN_0 \omega_{T_0} \omega_{T_1} \omega_L} E_{T_0} E_{T_1}^*.$$

One obtains from equation (54),

$$\frac{\partial}{\partial t}\left(|E_{T_0}|^2 + |E_{T_1}|^2 + |E_L|^2\right) = 0 \tag{55}$$

which implies conservation of the total wave energy density for the interacting waves.

Further, one obtains from equation (54),

$$-\frac{1}{\omega_{T_0}}\frac{\partial}{\partial t}|E_{T_0}|^2 = \frac{1}{\omega_{T_1}}\frac{\partial}{\partial t}|E_{T_1}|^2 = \frac{1}{\omega_L}\frac{\partial}{\partial t}|E_L|^2 \tag{56}$$

which are called Manley—Rowe [133] relations. The latter give the proportion of pump energy converted to each parametrically excited wave. Note that in equations (55) and (56), $\partial/\partial t$ refers to the slow variations in time.

One also obtains from equation (54), for the slow variations in time,

$$\begin{aligned}
\frac{\partial^2 E_{T_0}}{\partial t^2} &= -\frac{k_L^2\omega_p^4 E_{T_0}}{4e^2N_0^2\omega_{T_0}\omega_{T_1}}|E_L|^2 - \frac{k_L^2\omega_p^6 E_{T_0}}{4e^2N_0^2\omega_{T_0}\omega_{T_1}^2\omega_L}|E_{T_1}|^2\\
\frac{\partial^2 E_{T_1}}{\partial t^2} &= -\frac{k_L^2\omega_p^4 E_{T_1}}{4e^2N_0^2\omega_{T_0}\omega_{T_1}}|E_L|^2 + \frac{k_L^2\omega_p^6 E_{T_1}}{4e^2N_0^2\omega_{T_0}^2\omega_{T_1}\omega_L}|E_{T_0}|^2\\
\frac{\partial^2 E_L}{\partial t^2} &= -\frac{k_L^2\omega_p^6 E_L}{4e^2N_0^2\omega_{T_0}\omega_{T_1}^2\omega_L}|E_{T_1}|^2 + \frac{k_L^2\omega_p^6 E_L}{4e^2N_0^2\omega_L\omega_{T_0}^2\omega_{T_1}}|E_{T_0}|^2
\end{aligned} \tag{57}$$

which show the decay instability of the transverse electromagnetic wave (ω_{T_0}, k_{T_0}) into another transverse electromagnetic wave (ω_{T_1}, k_{T_1}) and a Langmuir wave (ω_L, k_L).

If one assumes that

$$|E_{T_0}| \gg |E_{T_1}|, |E_L|$$

then E_{T_0} can be taken to be constant. If further,

$$E_{T_1}, E_L \sim e^{\gamma t} \tag{58}$$

one obtains:

$$\gamma = \frac{k_L\omega_p^3}{2eN_0\omega_{T_0}\sqrt{\omega_{T_1}\omega_L}}|E_{T_0}|. \tag{59}$$

(*b*) *Nonlinear Resonant Interactions Between Two Transverse Electromagnetic Waves and an Ion-Acoustic Wave*

The nonlinear resonant interaction among two transverse electromagnetic waves and an ion-acoustic wave have been considered by Lashmore-Davies [16] and Shivamoggi [15].

Consider a plasma comprised of warm electrons and cold ions. The equations governing wavemotions in such a two-fluid plasma are:

$$\frac{\partial n_e}{\partial t} + N_0 \nabla \cdot \mathbf{V}_e = -\nabla \cdot (n_e \mathbf{V}_e) \tag{60}$$

$$\frac{\partial n_i}{\partial t} + N_0 \nabla \cdot \mathbf{V}_i = -\nabla \cdot (n_i \mathbf{V}_i) \tag{61}$$

$$\frac{\partial \mathbf{V}_e}{\partial t} + \frac{e\mathbf{E}}{m_e} + \frac{V_{T_e}^2}{N_0} \nabla n_e = -(\mathbf{V}_e \cdot \nabla)\mathbf{V}_e - \frac{e}{m_e c} \mathbf{V}_e \times \mathbf{B} + \frac{V_{T_e}^2}{N_0} n_e \nabla n_e \tag{62}$$

$$\frac{\partial \mathbf{V}_i}{\partial t} - \frac{e\mathbf{E}}{m_i} = -(\mathbf{V}_i \cdot \nabla)\mathbf{V}_i + \frac{e}{m_i c} \mathbf{V}_i \times \mathbf{B} \tag{63}$$

$$\frac{1}{c} \frac{\partial \mathbf{B}}{\partial t} + \nabla \times \mathbf{E} = 0 \tag{64}$$

$$\frac{\partial \mathbf{E}}{\partial t} - c\nabla \times \mathbf{B} + eN_0(\mathbf{V}_i - \mathbf{V}_e) = -e(n_i \mathbf{V}_i - n_e \mathbf{V}_e). \tag{65}$$

Whereas the left-hand sides in equations (60)—(65) represent the linearized problem, the right-hand sides represent the nonlinear effects.

Let us now consider wavemotions in the x-direction. One may separate equations (60)—(65) into those governing the transverse modes:

$$\frac{\partial V_{e_y}}{\partial t} + \frac{eE_y}{m_e} = \frac{e}{m_e c} V_{e_x} B_z - V_{e_x} \frac{\partial V_{e_y}}{\partial x} \tag{66}$$

$$\frac{1}{c} \frac{\partial B_z}{\partial t} + \frac{\partial E_y}{\partial x} = 0 \tag{67}$$

$$\frac{\partial E_y}{\partial t} + c \frac{\partial B_z}{\partial x} - eN_0 V_{e_y} = en_e V_{e_y} \tag{68}$$

where we have ignored the ion-response to the transverse waves, and have assumed that the polarization of the transverse modes is such that $E_z = B_y = V_{e_z} = 0$, and the equations governing the longitudinal modes:

$$\frac{\partial n_e}{\partial t} + N_0 \frac{\partial V_{e_x}}{\partial x} = -\frac{\partial}{\partial x} (n_e V_{e_x}) \tag{69}$$

$$\frac{\partial n_i}{\partial t} + N_0 \frac{\partial V_{i_x}}{\partial x} = -\frac{\partial}{\partial x} (n_i V_{i_x}) \tag{70}$$

$$\frac{\partial V_{e_x}}{\partial t} + \frac{eE_x}{m_e} + \frac{V_{T_e}^2}{N_0}\frac{\partial n_e}{\partial x} = -\frac{e}{m_e c} V_{e_y} B_z - V_{e_x}\frac{\partial V_{e_x}}{\partial x} + \frac{V_{T_e}^2}{N_0} n_e \frac{\partial n_e}{\partial x} \tag{71}$$

$$\frac{\partial V_{i_x}}{\partial t} - \frac{eE_x}{m_i} = \frac{e}{m_i c} V_{i_y} B_z - V_{i_x}\frac{\partial V_{i_x}}{\partial x} \tag{72}$$

$$\frac{\partial E_x}{\partial t} + eN_0(V_{i_x} - V_{e_x}) = -en_i V_{i_x} + en_e V_{e_x}. \tag{73}$$

Let us consider linearized transverse modes of the form $\exp(-ik_T x)$, and introduce, as before,

$$a_T = V_{e_y} + \frac{i\omega_T}{eN_0} E_y + \frac{ik_T c}{eN_0} B_z \tag{74}$$

where

$$\omega_T^2 = \omega_{p_e}^2 + k_T^2 c^2.$$

Using equations (66)—(68), one then obtains, as before:

$$\frac{\partial a_T}{\partial t} = i\omega_T a_T \tag{75}$$

so that a_T is a normal mode of the linearized problem associated with equations (66)—(68).

Similarly, let us consider linearized longitudinal modes of the form $\exp(-ik_s x)$, and introduce:

$$a_s = n_e + \frac{N_0\omega_s}{k_s V_{T_e}^2} V_{e_x} + \frac{N_0\omega_s}{k_s C_s^2} V_{i_x} - \frac{ik_s}{e} E_x \tag{76}$$

where,

$$C_s^2 \equiv \frac{m_e}{m_i} V_{T_e}^2, \quad \omega_s = k_s C_s.$$

Using equations (69)—(73), and assuming that $\omega_s \ll \omega_{p_i}$, one obtains:

$$\frac{\partial a_s}{\partial t} = i\omega_s a_s \tag{77}$$

so that a_s is a normal mode of the linearized ion-acoustic waves given by equations (69)—(73).

The transverse and the longitudinal modes which are uncoupled in the linearized problem become coupled in the nonlinear problem, as before. When the nonlinear terms are included, one obtains from equations (66)—(73):

$$\frac{\partial a_T}{\partial t} - i\omega_T a_T = \frac{e}{m_e c} V_{e_x} B_z - V_{e_x} \frac{\partial V_{e_y}}{\partial x} + \frac{i\omega_T}{N_0} n_e V_{e_y} \tag{78}$$

$$\begin{aligned}\frac{\partial a_s}{\partial t} - i\omega_s a_s = &- \frac{\partial}{\partial x}(n_e V_{e_x}) - \frac{N_0 \omega_s}{k_s V_{T_e}^2}\left(V_{e_x}\frac{\partial V_{e_x}}{\partial x}\right.\\ &\left.+ \frac{e}{m_e c} V_{e_y} B_z - \frac{V_{T_e}^2}{N_0} n_e \frac{\partial n_e}{\partial x}\right)\\ &+ \frac{N_0 \omega_s}{k_s C_s^2}\left(\frac{e}{m_i c} V_{i_y} B_z - V_{i_x}\frac{\partial V_{i_x}}{\partial x}\right)\\ &- \frac{ik_s}{e}(-en_i V_{i_x} + en_e V_{e_x}).\end{aligned} \tag{79}$$

Let us consider two transverse electromagnetic waves of the form $\exp[-i(k_{T_0}x - \omega_{T_0}t)]$ and $\exp[-i(k_{T_1}x - \omega_{T_1}t)]$ propagating in the x-direction, with

$$\omega_{T_0}^2 = \omega_{p_e}^2 + k_{T_0}^2 c^2, \quad \omega_{T_1}^2 = \omega_{p_e}^2 + k_{T_1}^2 c^2. \tag{80}$$

Due to the nonlinear resonant interaction between these two waves, let an ion-acoustic wave of the form $\exp[-i(k_s x - \omega_s t)]$, propagating in the x-direction, be excited such that

$$\omega_{T_0} - \omega_{T_1} = \omega_s, \quad k_{T_0} - k_{T_1} = k_s. \tag{81}$$

One obtains for the linearized problem:

$$\begin{aligned} V_e &= \frac{\omega_{p_e}^2}{2\omega_T^2} a_T \\ E_y &= -\frac{im_e}{e}\frac{\omega_{p_e}^2}{2\omega_T} a_T \\ B_z &= -\frac{im_e k_T c}{e}\frac{\omega_{p_e}^2}{2\omega_T^2} a_T \end{aligned} \tag{82}$$

$$a_T = a_T(t)e^{i(\omega_T t - k_T x)} + a_T^*(t)e^{-(\omega_T t - k_T x)}$$

and,

$$
\begin{aligned}
V_{e_x} &= V_{i_x} = \frac{C_s}{2N_0}\, a_s \\
n_e &= \frac{a_s}{2} \\
E_x &= \frac{i m_e k_s V_{T_e}^2}{2eN_0}\, a_s \\
a_s &= a_s(t)e^{i(\omega_s t - k_s x)} + a_s^*(t)e^{-i(\omega_s t - k_s x)}
\end{aligned}
\tag{83}
$$

where $a_T(t)$ and $a_s(t)$ are slowly-varying functions of time.

Using (82) and (83), equations (78) and (79) give, upon keeping only the resonant terms on their right-hand sides:

$$
\begin{aligned}
\frac{\partial a_{T_0}}{\partial t} - i\omega_{T_0} a_{T_0} &= \frac{i\omega_{p_e}^2 \omega_{T_0}}{4N_0 \omega_{T_1}^2}\, a_s a_{T_1} \\
\frac{\partial a_{T_1}}{\partial t} - i\omega_{T_1} a_{T_1} &= \frac{i\omega_{p_e}^2 \omega_{T_1}}{4N_0 \omega_{T_0}^2}\, a_s^* a_{T_0} \\
\frac{\partial a_s}{\partial t} - i\omega_s a_s &= \frac{iN_0 \omega_s \omega_{p_e}^4}{4\omega_{T_0} \omega_{T_1}^2 V_{T_e}^2}\, a_{T_0} a_{T_1}^*.
\end{aligned}
\tag{84}
$$

In terms of the electric-field components E_y (or E_T) and E_x (or E_L), equations (84) become:

$$
\begin{aligned}
\frac{\partial E_{T_0}}{\partial t} - i\omega_{T_0} E_{T_0} &= \frac{e\omega_{p_e}^2}{2m_e k_s V_{T_e}^2 \omega_{T_1}}\, E_s E_{T_1} \\
\frac{\partial E_{T_1}}{\partial t} - i\omega_{T_1} E_{T_1} &= -\frac{e\omega_{p_e}^2}{2m_e k_s V_{T_e}^2 \omega_{T_0}}\, E_s^* E_{T_0} \\
\frac{\partial E_s}{\partial t} - i\omega_s E_s &= -\frac{e m_s k_s}{2m_e \omega_{T_0} \omega_{T_1}}\, E_{T_0} E_{T_1}^*
\end{aligned}
\tag{85}
$$

from which one obtains for the slow variation in time:

$$\frac{\partial^2 E_{T_0}}{\partial t^2} = - \frac{e^2 \omega_{p_e}^4}{4 m_e^2 k_s^2 V_{T_e}^4 \omega_{T_0} \omega_{T_1}} |E_s|^2 - \frac{e^2 \omega_{p_e}^2 \omega_s}{4 m_e^2 V_{T_e}^2 \omega_{T_0} \omega_{T_1}^2} |E_{T_1}|^2$$

$$\frac{\partial^2 E_{T_1}}{\partial t^2} = - \frac{e^2 \omega_{p_e}^4}{4 m_e^2 k_s^2 V_{T_e}^2 \omega_{T_0} \omega_{T_1}} |E_s|^2 + \frac{e^2 \omega_{p_e}^2 \omega_s}{4 m_e^2 V_{T_e}^2 \omega_{T_0}^2 \omega_{T_1}} |E_{T_0}|^2 \tag{86}$$

$$\frac{\partial^2 E_s}{\partial t^2} = - \frac{e^2 \omega_{p_e}^2 \omega_s}{4 m_e^2 V_{T_e}^2 \omega_{T_0} \omega_{T_1}^2} |E_{T_1}|^2 + \frac{e^2 \omega_{p_e}^2 \omega_s}{4 m_e^2 V_{T_e}^2 \omega_{T_0}^2 \omega_{T_1}} |E_{T_0}|^2$$

which shows the decay instability of the transverse electromagnetic wave (ω_{T_0}, k_{T_0}) into another transverse electromagnetic wave (ω_{T_1}, k_{T_1}), and an ion-acoustic wave (ω_s, k_s).

If one assumes that

$$|E_{T_0}| \gg |E_{T_1}|,\ |E_s| \tag{87}$$

then E_{T_0} can be taken to be a constant. If, further,

$$E_{T_1},\ E_s \sim e^{\gamma t} \tag{88}$$

then equations (86) give for the growth rate:

$$\gamma = \frac{\omega_{p_e}^3}{2 N_0 e V_{T_e} \omega_{T_0}} \sqrt{\frac{\omega_s}{\omega_{T_1}}}\, |E_{T_0}|. \tag{89}$$

Comparison of (88) with (59) shows that the T-T-L wave-wave interaction (in (a)) has a lower growth rate than the T-T-S wave-wave interaction (in the present section).

The nonlinear resonant interaction between an electromagnetic wave, a Langmuir wave and an ion-acoustic wave has been considered by Sugihara [134]. A Lagrangian method to treat wave-wave interactions was developed by Dougherty [135], and was applied to special cases by Galloway and Kim [136].

(c) *Nonlinear Resonant Interactions Between Two Circularly-Polarized Waves and a Langmuir Wave*

Nonlinear resonant interactions between two circularly-polarized waves and a Langmuir wave, all propagating parallel to the direction of the applied magnetic field, have been considered by Sjolund and Stenflo [17], Chen and Lewak [18], Prasad [19], Lee [20], and Shivamoggi [21].

Consider a cold electron-plasma subjected to a constant magnetic field $\mathbf{B}_0 = B_0 \hat{\mathbf{i}}_x$ and embedded in a uniformly smeared-out background of immobile ions. The wave motion in such a plasma is governed by the following equations

$$\frac{\partial n}{\partial t} + \nabla \cdot (n\mathbf{V}) = 0 \tag{90}$$

$$\frac{\partial V}{\partial t} + (\mathbf{V} \cdot \nabla)\mathbf{V} = -\frac{e}{m}\left(\mathbf{E} + \frac{1}{c}\mathbf{V} \times \mathbf{B}\right) \tag{91}$$

$$\nabla \times \mathbf{E} = -\frac{1}{c}\frac{\partial \mathbf{B}}{\partial t} \tag{92}$$

$$\nabla \times \mathbf{B} = \frac{1}{c}\frac{\partial \mathbf{E}}{\partial t} - \frac{4\pi e}{c} n\mathbf{V}. \tag{93}$$

Let us now consider waves propagating in the x-direction. Let us write equations (90)—(93) in a form that exhibits the linearized parts and the non-linear parts separately, as before:

$$\frac{\partial n}{\partial t} + N_0 \frac{\partial v_x}{\partial x} = -\varepsilon \frac{\partial}{\partial x}(nv_x) \tag{94}$$

$$\frac{\partial v_x}{\partial t} + \frac{e}{m} E_x = -\varepsilon \left[v_x \frac{\partial v_x}{\partial x} + \frac{e}{mc}(v_y B_z - v_z B_y) \right] \tag{95}$$

$$\frac{\partial v_y}{\partial t} + \frac{e}{m} E_y + \frac{e}{mc} v_z B_0 = -\varepsilon \left[v_x \frac{\partial v_y}{\partial x} + \frac{e}{mc}(v_z B_x - v_x B_z) \right] \tag{96}$$

$$\frac{\partial v_z}{\partial t} + \frac{e}{m} E_z - \frac{e}{mc} v_y B_0 = -\varepsilon \left[v_x \frac{\partial v_z}{\partial x} + \frac{e}{mc}(v_x B_y - v_y B_x) \right] \tag{97}$$

$$\frac{\partial B_x}{\partial t} = 0 \tag{98}$$

$$\frac{\partial B_y}{\partial t} - c\frac{\partial E_z}{\partial x} = 0 \tag{99}$$

$$\frac{\partial B_z}{\partial t} + c\,\frac{\partial E_y}{\partial x} = 0 \tag{100}$$

$$\frac{\partial E_x}{\partial t} - 4\pi e N_0 v_x = \varepsilon \cdot 4\pi e n v_x \tag{101}$$

$$\frac{\partial E_y}{\partial t} + c\,\frac{\partial B_z}{\partial x} - 4\pi e N_0 v_y = \varepsilon \cdot 4\pi e n v_y \tag{102}$$

$$\frac{\partial E_z}{\partial t} - c\,\frac{\partial B_y}{\partial x} - 4\pi e N_0 v_z = \varepsilon \cdot 4\pi e n v_z \tag{103}$$

where the unsubscripted quantities refer to the perturbations characterized by the small parameter ε. Note again that the left hand sides in equations (94)—(103) represent the linear problem, and the right hand sides represent the nonlinear terms. Let us express the quantities **V**, **E** and **B** in the following form:

$$\begin{aligned} \mathbf{Q} &= Q_x \hat{\mathbf{i}}_x + \mathbf{Q}_\perp \\ \mathbf{Q}_\perp &\equiv \tfrac{1}{2}(Q_+ \hat{\mathbf{i}}_- + Q_- \hat{\mathbf{i}}_+) \\ Q_\pm &\equiv Q_y \pm i Q_z, \quad \hat{\mathbf{i}}_\pm \equiv \hat{\mathbf{i}}_y \pm i \hat{\mathbf{i}}_z. \end{aligned} \tag{104}$$

Note that this representation explicitly brings forth the presence of circularly-polarized waves in a magnetized plasma.

Using (104), equations (94)—(103) can be combined to give:

$$\frac{\partial^2 E_x}{\partial t^2} + \omega_p^2 E_x = \varepsilon \cdot 4\pi e \left[\frac{\partial}{\partial t}(n v_x) - N_0 \left\{ v_x \frac{\partial v_x}{\partial x} + \frac{ie}{2mc}(V_+ B_- - V_- B_+) \right\} \right] \tag{105}$$

$$\frac{\partial^2 E_\pm}{\partial t^2} - c^2 \frac{\partial^2 E_\pm}{\partial x^2} + \omega_p^2 E_\pm \mp 4\pi N_0 e i \Omega V_\pm = \varepsilon \cdot 4\pi e \left[\frac{\partial}{\partial t}(n V_\pm) - N_0 \left(v_x \frac{\partial V_\pm}{\partial x} \pm \frac{ie}{mc} v_x B_\pm \right) \right] \tag{106}$$

where,

$$\omega_p^2 \equiv \frac{4\pi N_0 e^2}{m}, \quad \Omega \equiv \frac{e B_0}{mc}.$$

It is obvious from equations (105) and (106) that three circularly-polarized waves propagating parallel to the applied magnetic field cannot interact. This result was also deduced by Krishan and Fukai [137] who used the quantum-field theory formalism for plasma wave interactions. Therefore, we shall consider the interaction of two circularly-polarized waves with a Langmuir wave.

Let us consider a right circularly-polarized electromagnetic wave and a left circularly-polarized electromagnetic wave with frequencies ω_1, ω_2 and wavenumbers k_1, k_2, respectively, propagating in the x-direction. One has

$$\begin{aligned} \omega_1^2 - k_1^2c^2 - \omega_p^2 \frac{\omega_1}{\omega_1 - \Omega} &= 0 \\ \omega_2^2 - k_2^2c^2 - \omega_p^2 \frac{\omega_2}{\omega_2 + \Omega} &= 0. \end{aligned} \tag{107}$$

Due to nonlinear interaction between these two waves, an electrostatic Langmuir wave with frequency ω_3 and wavenumber k_3, and propagating in the x-direction will be excited such that

$$\omega_1 - \omega_2 = \omega_3, \quad k_1 - k_2 = k_3. \tag{108}$$

In order to treat this three-wave interaction, let us use this time the method of multiple time-scales. Thus, let us seek solutions to equations (105) and (106) of the form:

$$Q(x, t; \varepsilon) = Q^{(0)}(x, t, \tilde{t}) + \varepsilon Q^{(1)}(x, t, \tilde{t}) + 0(\varepsilon^2) \tag{109}$$

where $\tilde{t} \equiv \varepsilon t$ is the slow-time scale characterizing the rate at which energy is exchanged between the three waves.

Using (109), equations (94)—(103) give

$$\begin{aligned} \mathbf{E}^{(0)} &= (a_3 e^{i\psi_3} + \overset{*}{a}_3 e^{-i\psi_3})\hat{\mathbf{i}}_x + (a_1 e^{i\psi_1} + \overset{*}{a}_1 e^{-i\psi_1})\hat{\mathbf{i}}_+ \\ &\quad + (a_2 e^{i\psi_2} + \overset{*}{a}_2 e^{-i\psi_2})\hat{\mathbf{i}}_- \\ \mathbf{B}^{(0)} &= -\frac{ick_1}{\omega_1}(a_1 e^{i\psi_1} + \overset{*}{a}_1 e^{-i\psi_1})\hat{\mathbf{i}}_+ \\ &\quad + \frac{ick_2}{\omega_2}(a_2 e^{i\psi_2} + \overset{*}{a}_2 e^{-i\psi_2})\hat{\mathbf{i}}_- \\ \mathbf{V}^{(0)} &= \frac{e}{im\omega_3}(a_3 e^{i\psi_3} - \overset{*}{a}_3 e^{-i\psi_3})\hat{\mathbf{i}}_x \\ &\quad + \left[\frac{ea_1 e^{i\psi_1}}{im(\omega_1 - \Omega)} - \frac{e\overset{*}{a}_1 e^{-i\psi_1}}{im(\omega_1 - \Omega)}\right]\hat{\mathbf{i}}_+ \\ &\quad + \left[\frac{ea_2 e^{i\psi_2}}{im(\omega_2 + \Omega)} - \frac{e\overset{*}{a}_2 e^{-i\psi_2}}{im(\omega_2 + \Omega)}\right]\hat{\mathbf{i}}_- \\ n^{(0)} &= \frac{eN_0 k_3}{im\omega_3^2}(a_3 e^{i\psi_3} - \overset{*}{a}_3 e^{-i\psi_3}) \end{aligned} \tag{110}$$

where,

$$\psi_s \equiv k_s x - \omega_s t; \quad a_s = a_s(\tilde{t}); \quad s = 1, 2, 3.$$

Using (110), equations (105) and (106) give for the $0(\varepsilon)$ problem:

$$\begin{aligned}
&\frac{\partial^2 E_x^{(1)}}{\partial t^2} + \omega_p^2 E_x^{(1)} \\
&\quad = \varepsilon \cdot 2i\omega_3 \left(\frac{\partial a_3}{\partial \tilde{t}} e^{i\psi_3} - \frac{\partial a_3}{\partial \tilde{t}} e^{-i\psi_3} \right) \\
&\quad - \varepsilon \frac{ie\omega_p^2}{2m} \left[\left\{ \frac{k_1}{\omega_1(\omega_2 + \Omega)} - \frac{k_2}{\omega_2(\omega_1 - \Omega)} \right\} a_1 a_2^* e^{i(\psi_1 - \psi_2)} \right. \\
&\quad \left. - \left\{ \frac{k_1}{\omega_1(\omega_2 + \Omega)} - \frac{k_2}{\omega_2(\omega_1 - \Omega)} \right\} a_1^* a_2 e^{-i(\psi_1 - \psi_2)} \right] \\
&\quad + \text{nonresonant terms} \qquad (111)
\end{aligned}$$

$$\begin{aligned}
&\frac{\partial^2 E_\pm^{(1)}}{\partial t^2} - c^2 \frac{\partial^2 E_\pm^{(1)}}{\partial x^2} + \omega_p^2 E_\pm^{(1)} \mp 4\pi N_0 ei\Omega V_\pm^{(1)} \\
&\quad = \varepsilon \cdot 2i\omega_{2,1} \left(\frac{\partial a_{2,1}}{\partial \tilde{t}} e^{i\psi_{2,1}} - \frac{\partial a_{2,1}^*}{\partial \tilde{t}} e^{-i\psi_{2,1}} \right) \\
&\quad + \text{nonresonant terms.} \qquad (112)
\end{aligned}$$

The removal of the secular terms in equations (111) and (112) requires

$$\frac{\partial a_{1,2}}{\partial \tilde{t}} \approx 0 \qquad (113a, b)$$

$$\frac{\partial a_3}{\partial \tilde{t}} = \frac{e\omega_p^2}{4m\omega_3} \left[\frac{k_1}{\omega_1(\omega_2 + \Omega)} - \frac{k_2}{\omega_2(\omega_1 - \Omega)} \right] a_1 a_2^*. \qquad (113c)$$

Equations (113) may be interpreted in the following manner: if two circularly-polarized waves with opposite polarizations and large amplitudes (so that the latter will change only negligibly, which is what equations (113a, b) imply) interact with one another, then equation (113c) implies that a Langmuir wave can be excited by this interaction. If the quantity in the rectangular brackets on the right hand side of equation (113c) is made positive by suitably choosing the parameters k_1, k_2 and Ω, then this Langmuir wave has the possibility of growing in time until it saturates due to collisional effects in the plasma.

Let us next consider two right-circularly-polarized electromagnetic waves with frequencies ω_1 and ω_2 and wavenumbers k_1 and k_2, respectively, propagating in the x-direction. Let these two waves engage in a nonlinear interaction with a Langmuir wave with frequency ω_3 and wavenumber k_3, and propagating in the x-direction such that

$$\omega_1 = \omega_3 - \omega_2, \quad k_1 = k_3 - k_2, \quad \omega_3 = \omega_p. \tag{114}$$

In order to treat this three-wave interaction, let us seek solutions to equations (105) and (106) of the form given in (109) again.

Using (109), equations (94)—(103) give:

$$\begin{aligned}
\mathbf{E}^{(0)} &= (a_3 e^{i\psi_3} + a_3^* e^{-i\psi_3})\hat{\mathbf{i}}_x + (a_1 e^{i\psi_1} + a_1^* e^{-i\psi_1} + a_2 e^{i\psi_2} + a_2^* e^{-i\psi_2})\hat{\mathbf{i}}_+ \\
\mathbf{B}^{(0)} &= -i\left[\frac{ck_1}{\omega_1}(a_1 e^{i\psi_1} + a_1^* e^{-i\psi_1}) + \frac{ck_2}{\omega_2}(a_2 e^{i\psi_2} + a_2^* e^{-i\psi_2})\right]\hat{\mathbf{i}}_+ \\
\mathbf{V}^{(0)} &= \frac{e}{im\omega_3}(a_3 e^{i\psi_3} - a_3^* e^{-i\psi_3})\hat{\mathbf{i}}_x \\
&\quad + \left[\frac{e}{im(\omega_1 - \Omega)}(a_1 e^{i\psi_1} - a_1^* e^{-i\psi_1}) \right. \\
&\quad \left. + \frac{e}{im(\omega_2 - \Omega)}(a_2 e^{i\psi_2} - a_2^* e^{-i\psi_2})\right]\hat{\mathbf{i}}_+ \\
n^{(0)} &= \frac{eN_0 k_3}{im\omega_3}(a_3 e^{i\psi_3} - a_3^* e^{-i\psi_3})
\end{aligned} \tag{115}$$

where,

$$\psi_s \equiv k_s x - \omega_s t; \quad a_s = a_s(\tilde{t}); \quad s = 1, 2, 3.$$

Using (105), equations (105) and (106) give for the $0(\varepsilon)$ problem:

$$\frac{\partial^2 E_x^{(1)}}{\partial t^2} + \omega_p^2 E_x^{(1)} = \varepsilon \cdot 2i\omega_3 \left(\frac{\partial a_3}{\partial \tilde{t}} e^{i\psi_3} - \frac{\partial a_3}{\partial \tilde{t}} e^{-i\psi_3}\right) + \text{nonresonant terms} \tag{116}$$

$$\frac{\partial^2 E_{-1}^{(1)}}{\partial t^2} - c^2 \frac{\partial^2 E_{-1}^{(1)}}{\partial x^2} + \omega_p^2 E_{-1}^{(1)} + 4\pi N_0 ei\Omega V_{-1}^{(1)}$$

$$= \varepsilon \cdot 2i\omega_1 \left(\frac{\partial a_1}{\partial \tilde{t}} e^{i\psi_1} - \frac{\partial a_1^*}{\partial \tilde{t}} e^{-i\psi_1} \right)$$

$$+ \varepsilon \frac{ie\omega_p^2}{m\omega_3} \left[\left\{ - \frac{k_3(\omega_3 - \omega_2)}{\omega_3(\omega_2 - \Omega)} + \frac{k_2}{\omega_2 - \Omega} - \frac{k_2}{\omega_2} \right\} \times \right.$$

$$\left. \times \left\{ a_3 a_2^* e^{i(\psi_3 - \psi_2)} - a_3^* a_2 e^{-i(\psi_3 - \psi_2)} \right\} \right]$$

$$+ \text{nonresonant terms} \tag{117}$$

$$\frac{\partial^2 E_{-2}^{(1)}}{\partial t^2} - c^2 \frac{\partial^2 E_{-2}^{(1)}}{\partial x^2} + \omega_p^2 E_{-2}^{(1)} + 4\pi N_0 ei\Omega V_{-2}^{(1)}$$

$$= \varepsilon \cdot 2i\omega_2 \left(\frac{\partial a_2}{\partial \tilde{t}} e^{i\psi_2} - \frac{\partial a_2^*}{\partial \tilde{t}} e^{-i\psi_2} \right)$$

$$+ \varepsilon \frac{ie\omega_p^2}{m\omega_3} \left[\left\{ - \frac{k_3(\omega_3 - \omega_1)}{\omega_3(\omega_1 - \Omega)} + \frac{k_1}{\omega_1 - \Omega} - \frac{k_1}{\omega_1} \right\} \times \right.$$

$$\left. \times \left\{ a_3 a_1^* e^{i(\psi_3 - \psi_1)} - a_3^* a_1 e^{-i(\psi_3 - \psi_1)} \right\} \right]$$

$$+ \text{nonresonant terms.} \tag{118}$$

The removal of the secular terms in equations (116)—(118) then requires

$$\frac{\partial a_1}{\partial \tilde{t}} = \frac{e\omega_p^2}{2m\omega_1\omega_3^2} \left[\frac{k_3\omega_1\omega_2 - k_2\omega_3\Omega}{\omega_2(\omega_2 - \Omega)} \right] a_3 a_2^* \tag{119a}$$

$$\frac{\partial a_2}{\partial \tilde{t}} = \frac{e\omega_p^2}{2m\omega_2\omega_3^2} \left[\frac{k_3\omega_1\omega_2 - k_1\omega_3\Omega}{\omega_1(\omega_1 - \Omega)} \right] a_3 a_1^* \tag{119b}$$

$$\frac{\partial a_3}{\partial \tilde{t}} \approx 0. \tag{119c}$$

Equation (119) may be interpreted as follows: if a Langmuir wave with a large amplitude interacts with two circularly-polarized waves with the same polarization and small amplitudes, then the amplitude of the Langmuir wave will change only negligibly with time (which is what equation (119c) implies). If the quantities in the rectangular brackets on the right hand sides of equations (119a) and (119b) are made positive by suitably choosing the parameters k_1, k_2 and Ω, then equations (119a) and (119b) indicate the possibility of these circularly-polarized waves growing in time, (this would be like the decay of a large-amplitude Langmuir wave into two circularly-polarized waves with the same polarization).

(*d*) *Nonlinear Resonant Interactions Between Three Extraordinary Waves*

Nonlinear resonant interaction between three extraordinary waves propagating perpendicular to the applied magnetic field was treated by Harker and Crawford [26], Das [27], and Shivamoggi [21].

Consider a warm electron-plasma subjected to a constant magnetic field $\mathbf{B}_0 = B_0\hat{\mathbf{i}}_z$, and embedded in a uniformly smeared-out background of ions. The wave motion in such a plasma is governed by the following equations:

$$\frac{\partial n}{\partial t} + \nabla \cdot (n\mathbf{V}) = 0 \tag{120}$$

$$\frac{\partial \mathbf{V}}{\partial t} + (\mathbf{V} \cdot \nabla)\mathbf{V} = -\frac{e}{m}\left(\mathbf{E} + \frac{1}{c}\mathbf{V} \times \mathbf{B}\right) - \frac{a^2}{n}\nabla n \tag{121}$$

$$\nabla \times \mathbf{E} = -\frac{1}{c}\frac{\partial \mathbf{B}}{\partial t} \tag{122}$$

$$\nabla \times \mathbf{B} = \frac{1}{c}\frac{\partial \mathbf{E}}{\partial t} - \frac{e}{c}n\mathbf{V} \tag{123}$$

where a is the speed of sound in the plasma.

Let us now consider wave motions in the x-direction. Equations (120)–(123) can be written in a form that exhibits the linearized parts and the nonlinear parts separately:

$$\frac{\partial n}{\partial t} + n_0\frac{\partial v_x}{\partial x} = -\varepsilon\frac{\partial}{\partial x}(nv_x) \tag{124}$$

$$\frac{\partial v_x}{\partial t} + \frac{eE_x}{m} + \Omega v_y + \frac{a^2}{n}\frac{\partial n}{\partial x}$$

$$= -\varepsilon\left(v_x\frac{\partial v_x}{\partial x} + \frac{e}{mc}v_yB_z - \frac{a^2}{n_0}n\frac{\partial n}{\partial x}\right) \tag{125}$$

$$\frac{\partial v_y}{\partial t} + \frac{eE_y}{m} - \Omega v_x = -\varepsilon\left(v_x\frac{\partial v_y}{\partial x} - \frac{e}{mc}v_xB_z\right) \tag{126}$$

$$\frac{\partial E_x}{\partial x} - en_0 v_x = \varepsilon(env_x) \tag{127}$$

$$\frac{\partial E_y}{\partial t} + c\,\frac{\partial B_z}{\partial x} - en_0 v_y = \varepsilon(env_y) \tag{128}$$

$$\frac{1}{c}\,\frac{\partial B_z}{\partial t} + \frac{\partial E_y}{\partial x} = 0 \tag{129}$$

where the subscript 0 denotes the unperturbed values, and the unsubscripted quantities refer to the perturbations characterized by the small parameter ε, and $\Omega \equiv eB_0/mc$. We have assumed the polarization of the extraordinary modes to be such that $E_z = B_y = v_z = 0$. Note that the left-hand sides in equations (124)–(129) represent the linear problem and the right-hand sides represent the nonlinear terms.

Let us consider linearized extraordinary modes of the form $e^{-ik_T x}$ and introduce

$$a_T = v_y + \frac{i\omega_T}{en_0}\left(\frac{\Omega^2}{D_T} + 1\right)E_y + \frac{ik_T c}{en_0}\left(\frac{\Omega^2}{D_T} + 1\right)B_z$$
$$+ \frac{a^2}{n_0}\,\frac{ik_T\Omega}{D_T}\,n + \frac{i\omega_T\Omega}{D_T}\,v_x - \frac{e}{m}\,\frac{\Omega}{D_T}\,E_x \tag{130}$$

where ω_T and k_T satisfy the linear dispersion relation:

$$(\omega_p^2 + k^2c^2 - \omega^2)(\omega_p^2 + k^2a^2 - \omega^2)$$
$$+ \Omega^2(k^2c^2 - \omega^2) = 0 \tag{131}$$

or

$$\omega_T^2 = \tfrac{1}{2}[k_T^2(a^2 + c^2) + 2\omega_p^2 + \Omega^2]$$
$$+ \tfrac{1}{2}[\{k_T^2(a^2 + c^2) + 2\omega_p^2 + \Omega^2\}^2$$
$$- 4(\omega_p^2 + k_T^2a^2)(\omega_p^2 + k_T^2c^2) - 4k_T^2c^2\Omega^2]^{1/2} \tag{132}$$

and

$$D_T \equiv \omega_p^2 + k_T^2a^2 - \omega_T^2, \quad \omega_p^2 = \frac{e^2 n_0}{m}\,.$$

Using equations (124)–(131), one obtains

$$\frac{\partial a_T}{\partial t} = i\omega_T a_T \tag{133}$$

so that a_T is a normal mode of the linearized problem associated with equations (124)–(129). Note that in the absence of the applied magnetic field (i.e., $\Omega \Rightarrow 0$), (130) represents the transverse electromagnetic waves (hence the justification for the use of a subscript T in (130)).

Similarly, let us consider another class of linearized extraordinary waves of the form $e^{-ik_L x}$ and introduce:

$$a_L = n + \frac{\omega_L n_0}{k_L a^2} v_x + \frac{ien_0}{mk_L a^2} E_x - \frac{in_0}{k_L a^2} \frac{D_L}{\Omega} v_y$$

$$+ \frac{\omega_L}{k_L a^2 \Omega} (\Omega^2 + D_L) E_y + \frac{c}{ea^2 \Omega} (\Omega^2 + D_L) B_z \tag{134}$$

where ω_L and k_L satisfy the linear dispersion relation (131):

$$\begin{aligned}\omega_L^2 = {} & \tfrac{1}{2}[k_L^2(a^2 + c^2) + 2\omega_p^2 + \Omega^2] \\ & - \tfrac{1}{2}[\{k_L^2(a^2 + c^2) + 2\omega_p^2 + \Omega^2\}^2 \\ & - 4(\omega_p^2 + k_L^2 a^2)(\omega_p^2 + k_L^2 c^2) - 4k_L^2 c^2 \Omega^2]^{1/2}\end{aligned} \tag{135}$$

and

$$D_L \equiv \omega_p^2 + k_L^2 a^2 - \omega_L^2.$$

Using equations (124)—(129), (131) and (134), one obtains:

$$\frac{\partial a_L}{\partial t} = i\omega_L a_L \tag{136}$$

so that a_L is another normal mode of the linearized problem associated with equations (124)—(129). Note that in the absence of the applied magnetic field (i.e., Ω, $D_L \Rightarrow 0$), (134) represents Langmuir waves (hence the justification for the use of a subscript L in (134)).

Note that in the method applied for the case without the applied magnetic field (in Section (a)), a_T is a linear combination of v_y, E_y, and B_z, and a_L is a linear combination of n, v_x, and E_x. In contrast, for the present case with the applied magnetic field, this method has to be modified in that a_L and a_T are both linear combination of all the variables in question — v_y, E_y, B_z, n, v_x, and E_x (Shivamoggi [21]).

The two sets of extraordinary waves given by (130) and (134) are uncoupled in the linearized problem. The coupling between the two sets of extraordinary modes given by (130) and (134) becomes effective in the nonlinear problem. When the nonlinear terms are included, one obtains, from equations (124)—(129):

$$\begin{aligned}\frac{\partial a_T}{\partial t} - i\omega_T a_T = {} & -v_x \frac{\partial v_y}{\partial x} + \frac{e}{mc} v_x B_z \\ & + \frac{i\omega_T}{n_0}\left(\frac{\Omega^2}{D_T} + 1\right) nv_y \\ & - \frac{a^2}{n_0} \frac{ik_T \Omega}{D_T} \frac{\partial}{\partial x}(nv_x) - \frac{i\omega_T \Omega}{D_T}\left(v_x \frac{\partial v_x}{\partial x}\right. \\ & \left. + \frac{e}{mc} v_y B_z - \frac{a^2}{n_0} n \frac{\partial n}{\partial x}\right) - \frac{e^2}{m} \frac{\Omega}{D_T} nv_x\end{aligned} \tag{137}$$

$$\frac{\partial a_L}{\partial t} - i\omega_L a_L = -\frac{\partial}{\partial x}(nv_x) - \frac{\omega_L n_0}{k_L a^2}\left(v_x \frac{\partial v_x}{\partial x}\right.$$

$$\left. + \frac{e}{mc} v_y B_z - \frac{a^2}{n_0} n \frac{\partial n}{\partial x}\right) + \frac{i\omega_p}{k_L a^2} nv_x$$

$$- \frac{in_0}{k_L a^2}\frac{D_L}{\Omega}\left(-v_x \frac{\partial v_y}{\partial x} + \frac{e}{mc} v_x B_z\right)$$

$$+ \frac{\omega_L}{k_L \Omega^2 a^2}(\Omega^2 + D_L) nv_y. \tag{138}$$

Let us consider the extraordinary waves of the form $\exp[-i(k_{T_0}x - \omega_{T_0}t)]$ and $\exp[-i(k_{T_1}x - \omega_{T_1}t)]$ propagating in the x-direction, with (ω_{T_0}, k_{T_0}) and (ω_{T_1}, k_{T_1}) related to each other according to (132). Due to nonlinear resonant interaction between these two extraordinary waves, let a third extraordinary wave of the form $\exp[-i(k_L x - \omega_L t)]$ propagating in the x-direction, be excited such that

$$\omega_{T_0} - \omega_{T_1} = \omega_L, \quad k_{T_0} - k_{T_1} = k_L \tag{139}$$

and (ω_L, k_L) are related to each other according to (135).

Now, one obtains for the linearized problem associated with equations (124)–(129):

$$\begin{aligned}
v_{y_T} &= \alpha a_T \\
E_{y_T} &= \left(\frac{i\omega_T e n_0}{k_T^2 c^2 - \omega_T^2}\right)\alpha a_T \\
B_{z_T} &= \left(\frac{ien_0 k_T c}{k_T^2 c^2 - \omega^2}\right)\alpha a_T \\
n_T &= \left(\frac{k_T n_0 \Omega}{iD_T}\right)\alpha a_T \\
v_{x_T} &= \left(\frac{\omega_T \Omega}{iD_T}\right)\alpha a_T \\
E_{x_T} &= \left(-\frac{en_0 \Omega}{D_T}\right)\alpha a_T \\
a_T &= a_T(t)e^{i(\omega_T t - k_T x)} + a_T^*(t)e^{-i(\omega_T t - k_T x)}
\end{aligned} \tag{140}$$

and,

$$
\begin{aligned}
v_{x_L} &= \beta a_L \\
n_L &= \left(\frac{k_L n_0}{\omega_L} \right) \beta a_L \\
E_{x_L} &= \left(- \frac{i e n_0}{\omega_L} \right) \beta a_L \\
v_{y_L} &= \left(\frac{i D_L}{\omega_L \Omega} \right) \beta a_L \\
E_{y_L} &= \left(\frac{D_L}{\Omega} \frac{e n_0 k_L c}{\omega_L^2 - k_L^2 c^2} \right) \beta a_L \\
B_{z_L} &= \left(\frac{D_L}{\omega_L \Omega} \frac{e n_0 k_L c}{\omega_L^2 - k_L^2 c^2} \right) \beta a_L \\
a_L &= a_L(t) e^{i(\omega_L t - k_L x)} + a_L^*(t) e^{-i(\omega_L t - k_L x)}
\end{aligned}
\tag{141}
$$

where,

$$
\alpha \equiv \left[1 + \frac{(\omega_T^2 + k_T^2 c^2) \left(\dfrac{\Omega^2}{D_T} + 1 \right)}{\omega_T^2 - k_T^2 c^2} + \frac{\Omega^2}{D_T} (D_T + 2\omega_T^2) \right]^{-1}
$$

$$
\beta \equiv \left[\frac{(\omega_p^2 + \omega_L^2 + k_L^2 a^2) n_0}{k_L \omega_L a^2} + \frac{n_0 D_L}{k_L \omega_L a^2 \Omega} \times \times \left\{ D_L + \frac{(\Omega^2 + D_L)(\omega_L^2 + k_L^2 c^2)}{\omega_L^2 - k_L^2 c^2} \right\} \right]^{-1}
$$

and $a_T(t)$ and $a_L(t)$ are slowly-varying functions of time.

Using (140) and (141), equations (137) and (138) give, upon keeping only the resonant terms on their right-hand sides:

$$\frac{\partial a_{T_0}}{\partial t} - i\omega_{T_0} a_{T_0} = \alpha_1 \beta \Bigg[ik_{T_1} + ik_L \frac{\omega_{T_1}}{\omega_L} \frac{D_L}{D_{T_1}}$$

$$+ ik_{T_1} \frac{\omega_p^2}{k_{T_1}^2 c^2 - \omega_{T_1}^2}$$

$$+ ik_L \frac{\omega_{T_1}}{\omega_L} \frac{D_L}{D_{T_1}} \frac{\omega_p^2}{k_L^2 c^2 - \omega_L^2}$$

$$+ i\omega_{T_0} \left(\frac{\Omega^2}{D_{T_0}} + 1 \right) \times$$

$$\times \left(\frac{k_L}{\omega_L} + \frac{k_{T_1}}{\omega_L} \frac{D_L}{D_{T_1}} \right)$$

$$+ \frac{ia^2 k_{T_0}^2 \Omega^2}{D_{T_0} D_{T_1}} \left(k_{T_1} + \frac{\omega_{T_1}}{\omega_L} k_L \right)$$

$$+ \frac{i\omega_{T_0} \Omega^2}{D_{T_0}} \Bigg\{ k_{T_0} \frac{\omega_{T_1}}{D_{T_1}}$$

$$+ k_L \frac{\omega_p^2}{\Omega^2} \frac{D_L}{\omega_L (k_L^2 c^2 - \omega_L^2)}$$

$$+ k_{T_1} \frac{\omega_p^2}{\Omega^2} \frac{D_L}{\omega_L (k_{T_1}^2 c^2 - \omega_{T_1}^2)}$$

$$- k_{T_0} k_{T_1} k_L \frac{a^2 n_0}{\omega_L D_{T_1}} \Bigg\}$$

$$+ i\omega_p^2 \frac{\Omega^2}{D_{T_0}} \left(\frac{k_{T_1}}{D_{T_1}} + k_L \frac{\omega_{T_1}}{\omega_L D_{T_1}} \right) \Bigg] a_{T_1} a_L^* \qquad (142a)$$

$$\frac{\partial a_{T_1}}{\partial t} - i\omega_{T_1} a_{T_1} = \alpha_0 \beta \left[ik_{T_0} + ik_L \frac{\omega_{T_0}}{\omega_L} \frac{D_L}{D_{T_0}} \right.$$

$$+ ik_{T_0} \frac{\omega_p^2}{k_{T_0}^2 c^2 - \omega_{T_0}^2}$$

$$+ ik_L \frac{\omega_{T_0}}{\omega_L} \frac{D_L}{D_{T_0}} \frac{\omega_p^2}{k_L^2 c^2 - \omega_L^2}$$

$$+ i\omega_{T_1} \left(\frac{\Omega^2}{D_{T_1}} + 1 \right) \times$$

$$\times \left(\frac{k_L}{\omega_L} - \frac{k_{T_0}}{\omega_L} \frac{D_L}{D_{T_0}} \right)$$

$$+ \frac{ia^2 k_{T_1}^2 \Omega^2}{D_{T_0} D_{T_1}} \left(k_{T_0} + \frac{\omega_{T_0}}{\omega_L} k_L \right)$$

$$+ \frac{i\omega_{T_1} \Omega^2}{D_{T_1}} \left\{ k_{T_1} \frac{\omega_{T_0}}{D_{T_0}} \right.$$

$$+ k_L \frac{\omega_p^2}{\Omega^2} \frac{D_L}{\omega_L (k_L^2 c^2 - \omega_L^2)}$$

$$- k_{T_0} \frac{\omega_p^2}{\Omega^2} \frac{D_L}{\omega_L (k_{T_0}^2 c^2 - \omega_{T_0}^2)}$$

$$\left. + k_{T_0} k_{T_1} k_L \frac{a^2 n_0}{\omega_L D_{T_0}} \right\}$$

$$\left. + i\omega_p^2 \frac{\Omega^2}{D_{T_1}} \left(\frac{k_{T_0}}{D_{T_0}} + k_L \frac{\omega_{T_0}}{\omega_L D_{T_0}} \right) \right] a_{T_0} a_L^* \qquad (142b)$$

$$\frac{\partial a_L}{\partial t} - i\omega_L a_L = \alpha_0 \alpha_1 \left[\frac{in_0\Omega^2}{k_L a^2 D_{T_0} D_{T_1}} (\omega_p^2 + k_L^2 a^2) \times \right.$$

$$\times (k_{T_0}\omega_{T_1} + k_{T_1}\omega_{T_0})$$

$$- \frac{i\omega_L n_0 /}{k_L a^2} \left\{ k_{T_0} \frac{\omega_p^2}{k_{T_0}^2 c^2 - \omega_{T_0}^2} \right.$$

$$- k_{T_1} \frac{\omega_p^2}{k_{T_1}^2 c^2 - \omega_{T_1}^2}$$

$$\left. + \frac{k_L \Omega^2}{D_{T_0} D_{T_1}} (-\omega_{T_0}\omega_{T_1} + k_{T_0} k_{T_1} n_0 a^2) \right\}$$

$$- \frac{in_0 D_L}{k_L a^2 D_{T_0} D_{T_1}} \left(-k_T \omega_{T_0} + k_{T_0}\omega_{T_1} \right.$$

$$- k_{T_1}\omega_{T_0} \frac{D_{T_1}}{D_{T_0}} \frac{\omega_p^2}{k_{T_1}^2 c^2 - \omega_{T_1}^2}$$

$$\left. - k_{T_0}\omega_{T_1} \frac{D_{T_0}}{D_{T_1}} \frac{\omega_p^2}{k_{T_0}^2 c^2 - \omega_{T_0}^2} \right)$$

$$\left. + \frac{i\omega_L n_0}{k_L a^2} (\Omega^2 + D_L) \left(\frac{k_{T_1}}{D_{T_1}} - \frac{k_{T_0}}{D_{T_0}} \right) \right] a_{T_0} a_{T_1}^* \qquad (142c)$$

where,

$$D_{T_0} \equiv D_T(\omega_{T_0}, k_{T_0}), \quad D_{T_1} \equiv D_T(\omega_{T_1}, k_{T_1})$$

$$\alpha_0 \equiv \alpha(\omega_{T_0}, k_{T_0}), \quad \alpha_1 \equiv \alpha(\omega_{T_1}, k_{T_1}).$$

In terms of a physical quantity like n_T and n_L (as given in (140) and (141), equations (142) can be written in the following form:

$$\frac{\partial n_{T_0}}{\partial \tilde{t}} \equiv \frac{\partial n_{T_0}}{\partial t} - i\omega_{T_0} n_{T_0} = -pn_{T_1} n_L \qquad (143a)$$

$$\frac{\partial n_{T_1}}{\partial \tilde{t}} \equiv \frac{\partial n_{T_1}}{\partial t} - i\omega_{T_1} n_{T_1} = qn_{T_0} n_L^* \qquad (143b)$$

$$\frac{\partial n_L}{\partial \tilde{t}} \equiv \frac{\partial n_L}{\partial t} - i\omega_L n_L = rn_{T_0} n_{T_1}^* \qquad (143c)$$

where p, q, and r are constants which are determined from the right-hand sides in equations (142).

One then obtains for the slow variations in time:

$$\frac{\partial^2 n_{T_0}}{\partial \tilde{t}^2} = -pqn_{T_0}|n_L|^2 - prn_{T_0}|n_{T_1}|^2 \tag{144a}$$

$$\frac{\partial^2 n_{T_1}}{\partial \tilde{t}^2} = -pqn_{T_1}|n_L|^2 + qrn_{T_1}|n_{T_0}|^2 \tag{144b}$$

$$\frac{\partial^2 n_L}{\partial \tilde{t}^2} = -prn_L|n_{T_1}|^2 + qrn_L|n_{T_0}|^2. \tag{144c}$$

If one assumes that

$$|n_{T_0}| \gg |n_{T_1}|, |n_L|$$

then n_{T_0} can be taken to be constant. If further,

$$n_{T_1}, n_L \sim e^{\sigma t} \tag{145a}$$

then equations (144) give for the growth rate:

$$\sigma = \sqrt{qr}\,|n_{T_0}|. \tag{145b}$$

Note that in the absence of the magnetic field (i.e., Ω, $D_L \Rightarrow 0$), equations (142) reduce to

$$\frac{\partial a_{T_0}}{\partial t} - i\omega_{T_0} a_{T_0} = \frac{ik_L^2 a^2 \omega_p^2 \omega_{T_0}}{4\omega_L^2 \omega_{T_1}^2 n_0} a_{T_1} a_L$$

$$\frac{\partial a_{T_1}}{\partial t} - i\omega_{T_1} a_{T_1} = \frac{ik_L^2 a^2 \omega_p^2 \omega_{T_1}}{4\omega_L^2 \omega_{T_0}^2 n_0} a_{T_0} a_L^*$$

$$\frac{\partial a_L}{\partial t} - i\omega_L a_L = \frac{in_0 \omega_L \omega_p^4}{4\omega_{T_0}^2 \omega_{T_1}^2} a_{T_0} a_{T_1}^*$$

where,

$$\omega_T^2 = \omega_p^2 + k_T^2 c^2, \quad \omega_L^2 = \omega_p^2 + k_L^2 a^2.$$

These are the same as those deduced in Section (a).

Nonlinear resonant interaction among three ordinary electromagnetic waves propagating perpendicular to the applied magnetic field was treated by Stenflo [22, 23], Krishan *et al.* [24], and Munoz and Dagach [25].

Wave-wave interactions can sometimes lead to what are called explosive instabilities (Coppi *et al.* [138], Dikasov *et al.* [139], Weiland and Wilhelmsson [132]). This arises in nonlinear interactions among waves of positive and negative energies (a space-charge wave on a beam travelling with a speed slower than that of the beam is said to have negative energy because excitation of such a wave results in lowering the total energy of the system). This leads to a possibility that all interacting waves reach infinite amplitude in a finite time. Experimental evidence of the existence of explosive instabilities was first obtained by Sugaya *et al.* [140]. In their experiment, Sugaya *et al.* [140] observed that the fast and slow space-charge waves of the helical electron beam which corresponded to the positive- and negative-energy waves, respectively, were amplified simultaneously by the nonlinear interaction with the beam.

(*e*) *Stimulated Wave-Scattering Phenomena*

One of the consequences of the interaction of an incident (pump) electromagnetic wave with a plasma is the parametric excitation of two plasma waves. If the latter are both purely electrostatic, they are eventually absorbed in the plasma and this decay process then leads to enhanced (or anomalous) absorption of the incident electromagnetic wave. If one of the excited plasma waves is electromagnetic, it can escape from the plasma and show up as enhanced (or stimulated) scattering of the incident electromagnetic wave. The latter process can be of two types according as the other excited plasma wave (stimulated Raman scattering) or an ion-acoustic wave (stimulated Brillouin scattering), (Bornatici [33], Shivamoggi [34]).

Stimulated Brillouin scattering and stimulated Raman scattering processes are of prime importance in connection with absorption and reflection of laser energy in a plasma. Intense laser radiation used to heat a plasma may undergo parametric decay into an ion-acoustic (or a Langmuir) wave and a backscattered light wave, resulting in reflection of the incident laser energy in the underdense, outer regions of the plasma, and thus diminishing absorption of the resonant layer. It happens that stimulated Brillouin scattering has a larger growth rate (see Ch. IV, Sec. (ii)), a lower threshold of excitation and a higher nonlinear saturation than stimulated Raman scattering, and can be so significant.

A Prototype for Parametric Instabilities. As a prototype for the parametric processes considered below, consider a system acted on by an oscillatory pump (of large magnitude) of the form

$$Z(t) = 2Z_0 \cos \omega_0 t \tag{146}$$

where Z_0 is taken to be a constant if one restricts consideration to initial stages of the ensuing instabilities in the system so that any depletion of the pump is then negligible. The pump induces a coupling between two natural modes of oscillation $X(t)$ and $Y(t)$ (with characteristic frequencies ω_1, ω_2, respectively) say, in the form (Nishikawa [32]):

$$\left(\frac{d^2}{dt^2} + \omega_1^2\right) X(t) = \lambda Y(t) Z(t) \tag{147}$$

$$\left(\frac{d^2}{dt^2} + \omega_2^2\right) Y(t) = \mu X(t) Z(t) \tag{148}$$

where λ and μ are coupling constants which are assumed to be such that $(\lambda\mu)$ is real and positive.

Upon Fourier transforming according to

$$Q(t) = \int_{-\infty}^{\infty} Q(\omega) e^{-i\omega t}\, d\omega \tag{149}$$

equations (147) and (148) give

$$(\omega^2 - \omega_1^2) X(\omega) = -\lambda Z_0 [Y(\omega - \omega_0) + Y(\omega + \omega_0)] \tag{150}$$

$$(\omega^2 - \omega_2^2) Y(\omega) = -\mu Z_0 [X(\omega - \omega_0) + X(\omega + \omega_0)]. \tag{151}$$

Consider a resonant situation with

$$\omega_0 \approx \omega_1 + \omega_2 \tag{152}$$

and consider equations (150) and (151) for $X(\omega)$, $Y(\omega \pm \omega_0)$ only, so that the dispersion relation follows

$$(\omega - \omega_1)(\omega - \omega_0 + \omega_2) + \frac{\lambda\mu Z_0^2}{4\omega_1\omega_2} = 0. \tag{153}$$

Putting,

$$\omega = \Omega + i\gamma \tag{154}$$

(153) gives

$$\gamma^2 \left(1 + \frac{\Delta^2}{4\gamma^2}\right) = \frac{\lambda\mu Z_0^2}{4\omega_1\omega_2} \tag{155}$$

where,

$$\Delta \equiv \omega_0 - \omega_1 - \omega_2.$$

The maximum growth rate occurs at perfect match ($\Delta = 0$), and is given by

$$\gamma_{\max} = \sqrt{\frac{\lambda\mu}{4\omega_1\omega_2}}\, Z_0. \tag{156}$$

Stimulated Raman Scattering. Consider a homogeneous plasma with a uniform background magnetic field B_0. (Experiments of Stamper *et al.* [109—111] and Diverglio *et al.* [112] showed that intense spontaneously generated magnetic fields are present in laser-produced plasmas. See Ch. VI, Sec. (ii) for a summary of these experiments. These magnetic fields are usually strong enough to modify the spectrum of electrostatic modes in the plasma, but not strong enough to influence the characteristics of propagation of the incident and scattered electromagnetic modes.) A large amplitude plane-polarized electromagnetic pump wave

$$\mathbf{E}_t = 2\mathbf{E}_t \cos(\mathbf{k}_t \cdot \mathbf{x} - \omega_t t) \tag{157}$$

with $\mathbf{E}_t$ parallel to $\mathbf{B}_0$, is incident on the plasma. The equilibrium state is comprised of electrons oscillating with velocity

$$\mathbf{V}_t = \frac{2e\mathbf{E}_t}{m_e\omega_0} \sin(\mathbf{k}_t \cdot \mathbf{x} - \omega_t t). \tag{158}$$

The ions remain stationary and make up a neutralizing background. Let us perturb this equilibrium and study the time-development of these perturbations using the linearized plasma equations. In the stimulated Raman scattering process in a plasma where only the electrons participate, the pump wave $(\omega_t, \mathbf{k}_t)$ decays into an electromagnetic wave $(\omega_{t'}, \mathbf{k}_{t'})$ and a Langmuir wave $(\omega_\ell, \mathbf{k}_\ell)$ with the constraints:

$$\omega_t = \omega_{t'} + \omega_\ell, \quad \mathbf{k}_t = \mathbf{k}_{t'} + \mathbf{k}_\ell. \tag{159}$$

One has for the Langmuir wave:

$$\frac{\partial n_\ell}{\partial t} + N_0 \nabla \cdot \mathbf{V}_\ell = 0 \tag{160}$$

$$\begin{aligned}\frac{\partial \mathbf{V}_\ell}{\partial t} &+ \frac{3V_{T_e}^2}{N_0} \nabla n_\ell + \frac{e}{m_e}\mathbf{E}_\ell + \frac{e}{m_e c}\mathbf{V}_\ell \times \mathbf{B}_0 \\ &= -(\mathbf{V}_t \cdot \nabla \mathbf{V}_{t'} + \mathbf{V}_{t'} \cdot \nabla \mathbf{V}_t) \\ &\quad - \frac{e}{m_e c}(\mathbf{V}_t \times \mathbf{B}_{t'} + \mathbf{V}_{t'} \times \mathbf{B}_t)\end{aligned} \tag{161}$$

where we have used the usual notations. Note from

$$\nabla \times \mathbf{E} = -\frac{1}{c}\frac{\partial \mathbf{B}}{\partial t} \tag{162}$$

that

$$\frac{e}{m_e c}\mathbf{B}_t = \nabla \times \mathbf{V}_t, \quad \frac{e}{m_e c}\mathbf{B}_{t'} = \nabla \times \mathbf{V}_{t'}. \tag{163}$$

Using (163), equation (161) becomes

$$\frac{\partial \mathbf{V}_\ell}{\partial t} + \frac{3V_{T_e}^2}{N_0}\nabla n_\ell + \frac{e}{m_e}\mathbf{E}_\ell + \frac{e}{m_e c}\mathbf{V}_\ell \times \mathbf{B}_0 = -\nabla(\mathbf{V}_t \cdot \mathbf{V}_{t'}). \tag{164}$$

Taking the divergence of equation (164), using

$$\nabla \cdot \mathbf{E}_\ell = -4\pi e n_\ell \tag{165}$$

and equation (160) and considering propagation perpendicular to $\mathbf{B}_0$, (i.e., $\mathbf{k}_\ell \cdot \mathbf{B}_0 = 0$) one obtains:

$$\left(\frac{\partial^2}{\partial t^2} + \omega_u^2\right) n_\ell = N_0 \nabla^2(\mathbf{V}_t \cdot \mathbf{V}_{t'}) \tag{166}$$

where,

$$\omega_u^2 = \omega_\ell^2 + \Omega_e^2, \quad \omega_\ell^2 = \omega_{p_e}^2 + 3k_\ell^2 V_{T_e}^2, \quad \Omega_e = \frac{eB_0}{m_e c}.$$

Next, from

$$\nabla^2 \mathbf{E}_{t'} - \frac{1}{c^2} \frac{\partial^2 \mathbf{E}_{t'}}{\partial t^2} = -\frac{4\pi e}{c^2} \frac{\partial}{\partial t} (n_\ell \mathbf{V}_{t'}) \tag{167}$$

one obtains for the scattered electromagnetic wave

$$\left(\frac{\partial^2}{\partial t^2} + \omega_{t'}^2 \right) \mathbf{E}_{t'} = 4\pi e \frac{\partial}{\partial t} (n_\ell \mathbf{V}_{t'}) \tag{168}$$

where,

$$\omega_{t'}^2 = \omega_{p_e}^2 + k_{t'}^2 c^2$$

and we have used the fact that the magnetic field $\mathbf{B}_0$ is not strong enough to influence the characteristics of propagation of the incident and scattered electromagnetic waves.

Noting,

$$\nabla^2 (\mathbf{V}_t \cdot \mathbf{V}_{t'}) = -\frac{e^2 k_\ell^2}{m_e^2 \omega_t \omega_{t'}} \mathbf{E}_t \cdot \mathbf{E}_{t'}$$

$$\frac{\partial}{\partial t} (n_\ell \mathbf{V}_t) = -\frac{e\omega_{t'}}{m_e \omega_t} n_\ell \mathbf{E}_t \tag{169}$$

equations (166) and (168) become

$$\left(\frac{\partial^2}{\partial t^2} + \omega_u^2 \right) n_\ell = \lambda_R E_t E_{t'} \tag{170}$$

$$\left(\frac{\partial^2}{\partial t^2} + \omega_{t'}^2 \right) E_{t'} = \mu_R E_t n_\ell \tag{171}$$

where,

$$\lambda_R \equiv -\frac{\omega_{p_e}^2}{\omega_t \omega_{t'}} \frac{k_\ell^2}{4\pi m_e} (\hat{\mathbf{e}}_t \cdot \hat{\mathbf{e}}_{t'})$$

$$\mu_R \equiv -\frac{\omega_{t'}}{\omega_t} \frac{\omega_{p_e}^2}{N_0} (\hat{\mathbf{e}}_t \cdot \hat{\mathbf{e}}_{t'})$$

and $\hat{\mathbf{e}}_t$ and $\hat{\mathbf{e}}_{t'}$ are the directions of polarization of $\mathbf{E}_t$ and $\mathbf{E}_{t'}$.

Equations (170) and (171) are of the same form as equations (147) and (148), so that the maximum growth rate for stimulated Raman scattering is given from (100) as

$$\gamma_R = \frac{\omega_{p_e}^2 k_\ell |\hat{\mathbf{e}}_t \cdot \hat{\mathbf{e}}_{t'}|}{4\omega_t \sqrt{\pi N_0 m_e \omega_{t'} \omega_u}} E_t \tag{172}$$

observe that the growth rate is reduced in the presence of a background magnetic field.

Unlike simulated Brillouin scattering, which involves an ion-acoustic wave with little dispersion and therefore satisfies easy matching conditions, stimulated Raman scattering requires a long-scale-length underdense plasma in order to provide phase matching and growth for the highly dispersive electron plasma wave. As a consequence, stimulated Raman scattering has not been extensively studied in current laser-target interaction experiments in which density-gradient scale lengths, L_n, are short. A linear magnetically confined plasma, on the other hand, offers the prospect for providing large L_n and was used by Watt *et al.* [140] who reported observations of stimulated Raman backscatter in an underdense hydrogen plasma. Stimulated Raman backscatter is characterized by an exponential backscattered-intensity dependence on the incident intensity and a backscattered frequency which is shifted to a lower frequency than the incident frequency by the electron plasma frequency. The overall backscattered intensity as a function of incident intensity was measured and shown in Figure (4.7a). The latter clearly indicates the exponential behavior of the backscattered radiation. Since both stimulated Raman and stimulated Brillouin processes may be present in this data, Watt *et al.*, spectrally dispersed the backscattered radiation and the resulting spectra is shown in Figure (4.7b). The data in Figure (4.7b) is correct for the low-density, cold plasma in the theta pinch but in later, more highly resolved spectral experiments in this device, a more complicated shape was seen (R. G. Watt, personal communication, 1987). In particular, the curve drawn in Figure (4.7b) for the data at less than 10.9 μm was later examined in more detail and was found not to fall as linearly as indicated. The data point above the curve is real, with later spectrum showing a pronounced dual lobe nature. Offenberger *et al.* [37] reported spectrally and temporally resolved measurements of growth and saturation of stimulated Raman backscatter from long-scale-length, magnetically confined, underdense hydrogen plasma column for incident CO_2 laser intensities up to 2.5×10^{11} w/cm^2. Typical single-shot, time, integrated spectral measurement is shown in Figure 4.8. When there was predominantly one backscatter pulse, a single spectral peak was observed.

Stimulated Brillouin Scattering. Here the pump wave $(\omega_t, \mathbf{k}_t)$ decays into an electromagnetic wave $(\omega_{t'}, \mathbf{k}_{t'})$ and an ion-acoustic wave $(\omega_s, \mathbf{k}_s)$, with the constraints:

$$\omega_t = \omega_{t'} + \omega_s, \quad \mathbf{k}_t = \mathbf{k}_{t'} + \mathbf{k}_s. \tag{173}$$

Here both electrons and ions participate in the motion of an ion-acoustic wave.

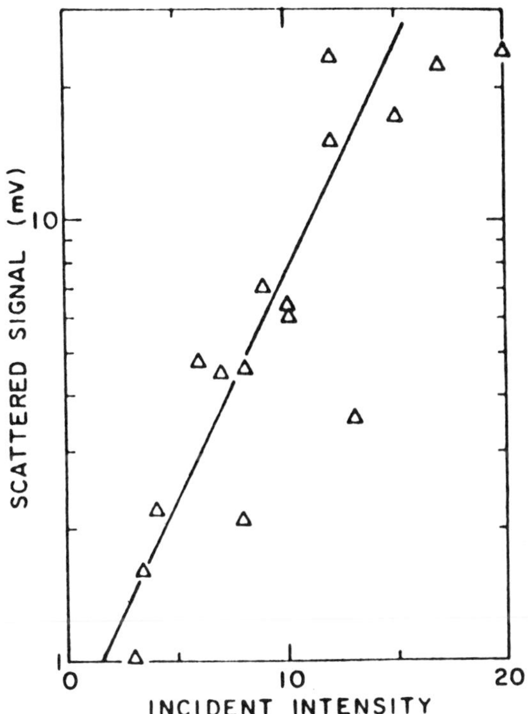

Figure 4.7a. Measured backscattered-signal amplitude as a function of incident CO_2 intensity. (Due to Watt *et al.* [40], by courtesy of The American Physical Society.)

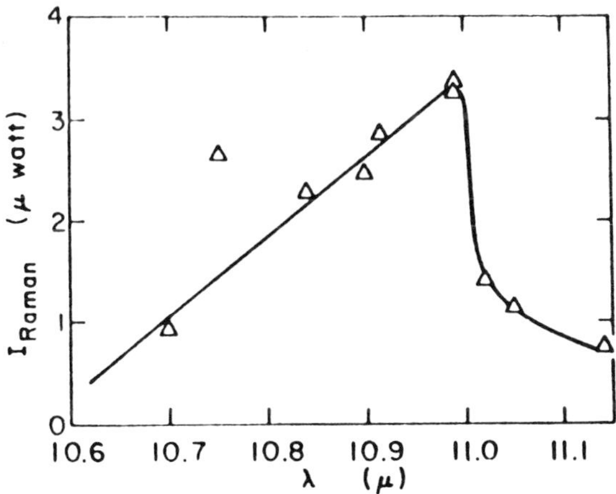

Figure 4.7b. Backscattered spectrum as a function of wavelength. (Due to Watt *et al.* [40], by courtesy of The American Physical Society.)

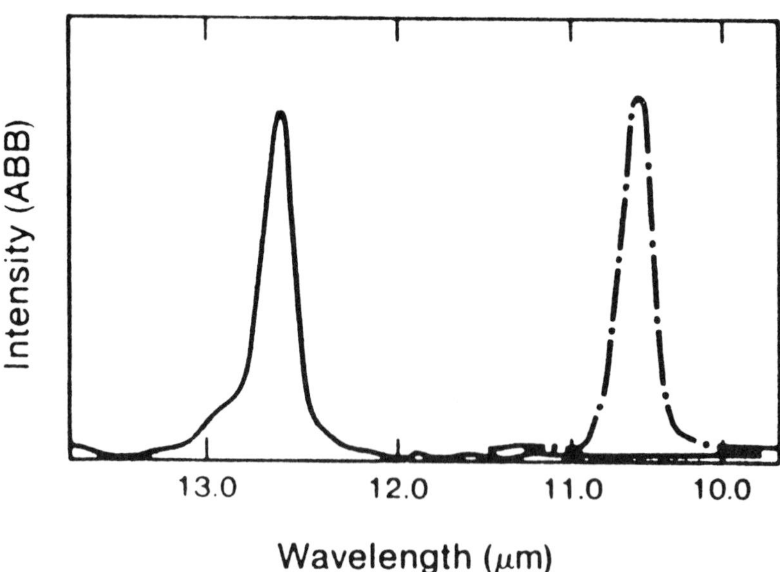

Figure 4.8. Raman backscatter spectra (solid lines) and CO_2 input reference spectra (dashed lines) for the case of predominantly a single backscatter spike. (Due to Offenberger *et al.* [37], by courtesy of The American Physical Society.)

One has for the electrons moving in the latter:

$$\frac{\partial n_{e_s}}{\partial t} + N_0 \nabla \cdot \mathbf{V}_{e_s} = 0 \tag{174}$$

$$\begin{aligned} \frac{V_{T_e}^2}{N_0} \nabla n_{e_s} + \frac{e}{m_e} \mathbf{E}_s + \frac{e}{m_e c} \mathbf{V}_{e_s} \times \mathbf{B}_0 \\ = -(\mathbf{V}_t \cdot \nabla \mathbf{V}_{t'} + \mathbf{V}_{t'} \cdot \nabla \mathbf{V}_t) \\ - \frac{e}{m_e c} (\mathbf{V}_t \times \mathbf{B}_{t'} + \mathbf{V}_{t'} \times \mathbf{B}_t) = -\nabla(\mathbf{V}_t \cdot \mathbf{V}_{t'}) \end{aligned} \tag{175}$$

where we have ignored the electron inertia, and have assumed that the electrons respond isothermally to the ion-acoustic wave. One has for the ions moving in the latter:

$$\frac{\partial n_{i_s}}{\partial t} + N_0 \nabla \cdot \mathbf{V}_{i_s} = 0 \tag{176}$$

$$\frac{\partial \mathbf{V}_{i_s}}{\partial t} + \frac{3 V_{T_i}^2}{N_0} \nabla n_{i_s} - \frac{e}{m_i} \mathbf{E}_s - \frac{e}{m_i c} \mathbf{V}_{i_s} \times \mathbf{B}_0 = 0 \tag{177}$$

with

$$\nabla \cdot \mathbf{E}_s = 4\pi e (n_{i_s} - n_{e_s}). \tag{178}$$

Here V_{T_i} is the thermal speed of the ions.

Taking the divergence of equation (175), using equations (174) and (178), and considering propagation perpendicular to $\mathbf{B}_0$, (i.e., $\mathbf{k}_s \cdot \mathbf{B}_0 = 0$), one obtains

$$n_{e_s} = \frac{1}{1 + k_s^2\lambda_{D_e}^2 + \dfrac{\Omega_e^2}{\omega_{p_e}^2}} \left[n_{i_s} + \frac{N_0}{\omega_{p_e}^2} \nabla^2(\mathbf{V}_t \cdot \mathbf{V}_{t'}) \right] \tag{179}$$

where

$$\lambda_{D_e}^2 \equiv \frac{V_{T_e}^2}{\omega_{p_e}^2}.$$

Taking the divergence of equation (177), and using equations (176), (178) and (179), one obtains

$$\left(\frac{\partial^2}{\partial t^2} + \omega_s^2 + \Omega_i^2 \right) n_{i_s} = \frac{m_e}{m_i} \frac{N_0 \nabla^2(\mathbf{V}_t \cdot \mathbf{V}_{t'})}{1 + k_s^2\lambda_{D_e}^2 + \Omega_e^2/\omega_{p_e}^2} \tag{180}$$

where,

$$\omega_s^2 \equiv \frac{k_s^2 C_s^2}{1 + k_s^2\lambda_{D_e}^2 + \Omega_e^2/\omega_{p_e}^2} + 3k_s^2 V_{T_i}^2, \quad C_s^2 \equiv \frac{KT_e}{m_i}.$$

Using

$$\nabla^2 \mathbf{E}_{t'} - \frac{1}{c^2} \frac{\partial^2 \mathbf{E}_{t'}}{\partial t^2} = - \frac{4\pi e}{c^2} \frac{\partial}{\partial t} (n_{e_s} \mathbf{V}_t) \tag{181}$$

and (179), one obtains for the scattered electromagnetic wave,

$$\left(\frac{\partial^2}{\partial t^2} + \omega_{t'}^2 \right) \mathbf{E}_{t'} = \frac{4\pi e}{1 + k_s^2\lambda_{D_e}^2 + \Omega_e^2/\omega_{p_e}^2} \frac{\partial}{\partial t} (\mathbf{V}_t n_{i_s}). \tag{182}$$

Using (189), equations (180) and (182) can be written as

$$\left(\frac{\partial^2}{\partial t^2} + \omega_s^2 + \Omega_i^2 \right) n_{i_s} = \lambda_B E_t E_{t'} \tag{183}$$

$$\left(\frac{\partial^2}{\partial t^2} + \omega_{t'}^2 \right) E_{t'} = \mu_B E_t n_{i_s} \tag{184}$$

where

$$\lambda_B \equiv - \frac{\dfrac{\omega_{p_e}^2}{\omega_t \omega_{t'}} \dfrac{k_s^2}{4\pi m_i}}{1 + k_s\lambda_{D_e}^2 + \Omega_e^2/\omega_{p_e}^2} (\hat{\mathbf{e}}_t \cdot \hat{\mathbf{e}}_{t'})$$

$$\mu_B \equiv -\frac{\dfrac{\omega_{t'}}{\omega_t}\dfrac{\omega_{p_e}^2}{N_0}}{1 + k_s\lambda_{D_e}^2 + \Omega_e^2/\omega_{p_e}^2}\,(\hat{\mathbf{e}}_t \cdot \hat{\mathbf{e}}_{t'}).$$

Equations (183) and (184) are again of the same form as equations (147) and (148), so that the maximum growth rate for stimulated Brillouin scattering is given from (156) as

$$\gamma_B = \frac{\dfrac{\omega_{p_e}^2 k_s}{\omega_t\sqrt{4\pi m_i N_0}}\,|\hat{\mathbf{e}}_t \cdot \hat{\mathbf{e}}_{t'}|}{2(1 + k_s^2\lambda_{D_e}^2 + \Omega_e^2/\omega_{p_e}^2)\sqrt{\omega_{t'}(\omega_s^2 + \Omega_i^2)^{1/2}}} \tag{185}$$

observe that the growth rate is again reduced in the presence of a background magnetic field.

Ripin *et al.* [35] presented time-resolved and time-integrated measurements of the energy, spectra and other characteristics such as growth rate and threshold values of the backscattered laser radiation from a laser-produced plasma. The rapid rise and directional properties of the reflected radiation observed by them were consistent with the stimulated Brillouin-backscattering instability. The total energy of the back-scattered radiation increases exponentially with incident laser energy (see Figure 4.9) which is characteristic of the linear

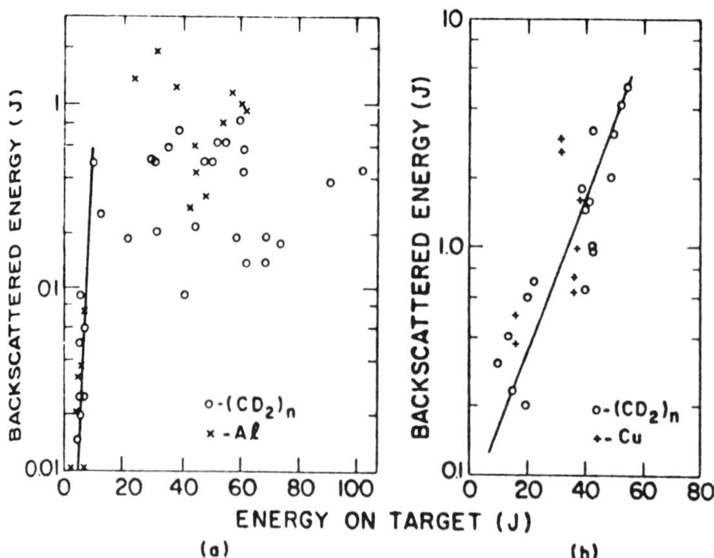

Figure 4.9. Laser energy backscattered through the focusing lens near 1.06 μm versus incident energy for (a) deuterated polyethylene and aluminum slab targets, with an $f/14$ focusing lens and 900-psec (full width at half-maximum) laser pulses, and (b) deuterated polyethylene and copper slab targets, with an $f/1.9$ focusing lens and 900-psec (full width at half-maximum) laser pulses. In this plot the specular-reflection contribution (1% of the incident energy) is subtracted from the back-reflected energy. (Due to Ripin *et al.* [35], by courtesy of The American Physical Society.)

behavior of stimulated Brillouin backscattering. This latter result was also observed by Offenberger *et al.* [36] and Grek *et al.* [39] who gave a further evidence of the existence of stimulated Brillouin scattering by examining the shape, power dependence and frequency shift of the backscattered spectrum. The spectrum of the backscattered radiation from an underdense plasma irradiated with CO_2 laser intensities less than 10^{13} w/cm^2 found by Ng *et al.* [41] is given in Figure 4.10, which shows a clear red shift characteristic of Brillouin scattering.

Turechek and Chen [38] confirmed observations of stimulated Brillouin scattering by — (a) checking the frequency shift, (b) checking the magnitude of the threshold, and (c) detecting the heat-frequency field spectroscopically. Ng *et al.* [41] found that strong stimulated Brillouin—Compton scattering saturates when long-scale-length underdense plasma was irradiated with focused CO_2-laser intensities less than 10^{13} w/cm^2. The magnitude of backward power as a function of incident power is shown in Figure 4.11. Saturation of the stimulated Brillouin scatter signal is a significant departure from the pure exponential growth observed with lesser laser intensity in the same plasma.

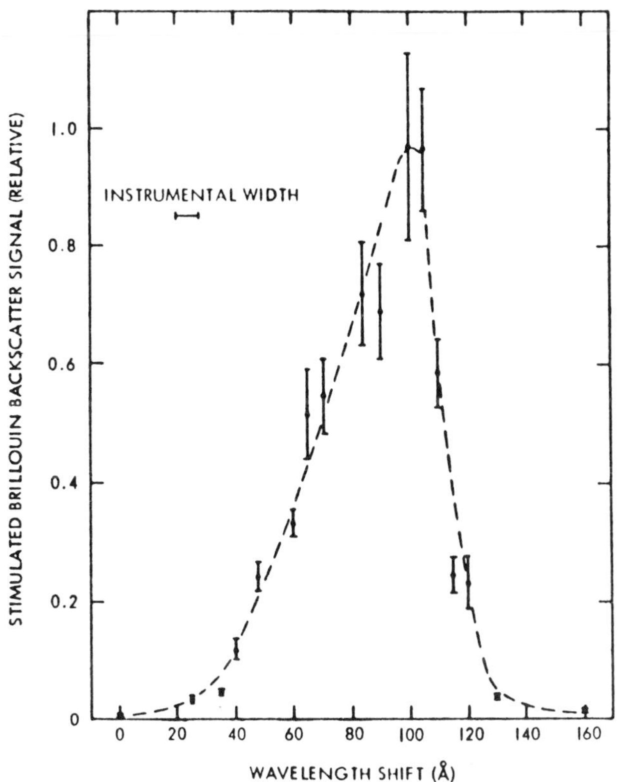

Figure 4.10. Spectrum of backscattered radiation showing the red-shifted Brillouin—Compton component. (Due to Ng *et al.* [41], by courtesy of The American Physical Society.)

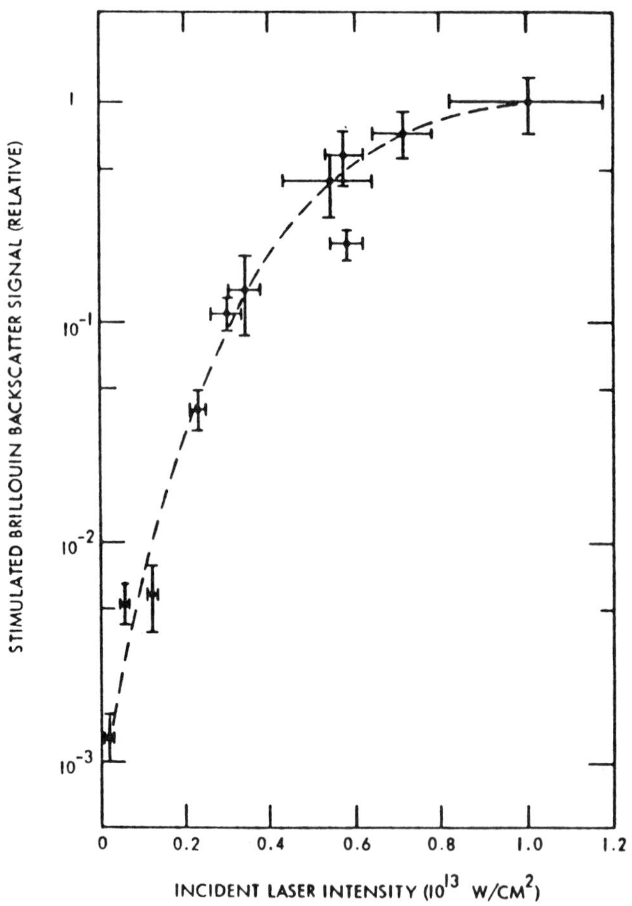

Figure 4.11. Stimulated Brillouin-backscatter signal as a function of incident CO_2-laser intensity. Unity in the ordinate corresponds to 30% time-averaged reflectivity. (Due to Ng *et al.* [41], by courtesy of The American Physical Society.)

(*f*) *Oscillating Two-Stream Instability*

A coupled-mode approach to the situation in which high — and low — frequency longitudinal oscillations in a plasma are excited simultaneously by some driving frequency ω_0 shows that there are two basic types of instability: one of the decay type considered in the previous sections, and a second in which the low-frequency oscillation is a zero-frequency oscillation or a purely-growing mode. This instability is often referred to as oscillating two-stream instability because under certain circumstances it can be regarded as being analogous to the linear DC two-stream instability but with the streaming motion now caused by the oscillatory driving field. Such a situation arises when an electromagnetic wave with frequency ω_0 interacts with a plasma with $\omega_0 \approx \omega_p$. This interaction leads to the decay of the transverse wave into an ion-acoustic wave and a

Langmuir wave. The frequency of this ion-acoustic wave will then be very much less than the frequencies of the Langmuir and pump waves.

Consider an electron set into motion by an electromagnetic wave with a nonuniform amplitude. One has

$$m_e \frac{d\mathbf{V}_e}{dt} = -e\left[\mathbf{E}_0(\mathbf{x}) + \frac{1}{c}\mathbf{V}_e \times \mathbf{B}(\mathbf{x})\right] \tag{186}$$

where

$$\mathbf{E}(\mathbf{x}) = \overline{\mathbf{E}}_0(\mathbf{x}) \cos \omega_0 t.$$

The electron performs rapid oscillations in the rapidly varying electric field about an oscillation center $\tilde{\mathbf{x}}_0$ which is also moving. Let us write

$$\mathbf{x} = \tilde{\mathbf{x}}_0 + \mathbf{x}_0, \quad |\mathbf{x}_0| \ll |\tilde{\mathbf{x}}_0|. \tag{187}$$

Then, to first order, one has

$$m_e \frac{d\mathbf{V}_{e_0}}{dt} = -e\mathbf{E}_0(\tilde{\mathbf{x}}_0) \tag{188}$$

from which,

$$\mathbf{V}_{e_0} = -\frac{e}{m_e\omega_0}\overline{\mathbf{E}}_0(\tilde{\mathbf{x}}) \sin \omega_0 t \tag{189}$$

and

$$\mathbf{x}_{e_0} = \frac{e}{m_e\omega_0^2}\overline{\mathbf{E}}_0(\tilde{\mathbf{x}}_0) \cos \omega_0 t. \tag{190}$$

Let us write,

$$\begin{aligned} \mathbf{E}_0(\mathbf{x}) &= \mathbf{E}_0(\mathbf{x}_0) + (\mathbf{x}_{e_1} \cdot \nabla)\mathbf{E}_0|_{\mathbf{x}=\mathbf{x}_0} + \cdots \\ \mathbf{B}_1 &= -\frac{1}{\omega_0}\nabla \times \overline{\mathbf{E}}_0|_{\mathbf{x}=\mathbf{x}_0} \sin \omega_0 t. \end{aligned} \tag{191}$$

One now has to second order:

$$m_e \frac{d\mathbf{V}_{e_1}}{dt} = -e\,[(\mathbf{x}_{e_1} \cdot \nabla)\mathbf{E}_0 + \mathbf{V}_{e_1} \times \mathbf{B}_1]. \tag{192}$$

Averaging equations (192) over the fast time scale (i.e., ω_0^{-1}), and using (189) and (190), one obtains

$$\begin{aligned} m_e\left\langle \frac{d\mathbf{V}_{e_1}}{dt}\right\rangle &= -\frac{e^2}{m_e\omega_0^2}\frac{1}{2}[(\overline{\mathbf{E}}_0 \cdot \nabla)\overline{\mathbf{E}}_0 + \overline{\mathbf{E}}_0 \times (\nabla \times \overline{\mathbf{E}}_0)] \\ &= -\frac{1}{4}\frac{e^2}{m_e\omega_0^2}\nabla(\overline{\mathbf{E}}_0^2) \end{aligned}$$

or

$$m_e N_0 \left\langle \frac{\mathrm{d}\mathbf{V}_{e_1}}{\mathrm{d}t} \right\rangle = -\frac{\omega_p^2}{\omega_0^2} \nabla \frac{\langle \mathbf{E}_0^2 \rangle}{8\pi} . \tag{193}$$

The origin of the effective nonlinear force represented by the right hand side is as follows: when the wave-amplitude is nonuniform, the electrons will pile up in regions of small wave-amplitude, and a force is needed to overcome the space charge. Although this ponderomotive force acts mainly on the electrons, this force is ultimately transmitted to the ions through the ambipolar field, since it is a low-frequency effect. The ponderomotive force thus behaves like a pressure acting to force the plasma out of high-field regions. A direct effect of the ponderomotive force is the self-focussing of laser light in a plasma (Chen [141]). This force moves the plasma out of a laser beam, so that ω_p is lower and the dielectric constant is higher inside the beam than outside. The plasma then acts as a convex lens focussing the laser beam to a smaller diameter.

It may be mentioned that the concept of ponderomotive force is not so clear in the case of a magnetized plasma (Bezzerides *et al.* [142], Johnston *et al.* [143], Belkov and Tsytovich [144], Kono *et al.* [145, 146], Festeau-Barrioz and Weibel [147], Tskhakaya [148], Karpman *et al.* [149, 150], Statham and ter Haar [151]). Nevertheless, it is useful to introduce a pseudopotential ϕ_p such that its gradient in the direction parallel to the magnetic field reproduces the time-independent force in that direction. Such a description is useful when the ponderomotive interaction is conservative, namely, when the particles leave the interaction region with their initial total energy. However, a straightforward description displays a singularity corresponding to the gyroresonance. In order to derive a proper description of the ponderomotive force near gyroresonance, one needs to take into account the fact that the interaction is not conservative near gyroresonance because of particle acceleration (Lamb *et al.* [152]).

One has for the ion-motion in the x-direction,

$$m_i N_0 \frac{\partial V_{i_1}}{\partial t} = F_{NL} \tag{194}$$

$$\frac{\partial n_{i_1}}{\partial t} + N_0 \frac{\partial V_{i_1}}{\partial x} = 0 \tag{195}$$

where,

$$\begin{aligned} F_{NL} &= -\frac{\omega_p^2}{\omega_0^2} \frac{\partial}{\partial x} \frac{\langle (\mathbf{E}_0 + \mathbf{E}_1)^2 \rangle}{8\pi} \\ &\approx -\frac{\omega_p^2}{\omega_0^2} \frac{\partial}{\partial x} \frac{\langle 2\mathbf{E}_0 \cdot \mathbf{E}_1 \rangle}{8\pi} \end{aligned} \tag{196}$$

$\mathbf{E}_1$ being the space-charge field.

Upon Fourier analyzing in space, equations (194) and (195) give:

$$\frac{\partial^2 n_{i_1}}{\partial t^2} + \frac{ik}{m_i} F_{NL} = 0. \tag{197}$$

Next, one has for the electrons

$$m_e \left(\frac{\partial V_e}{\partial t} + V_e \frac{\partial V_e}{\partial x} \right) = -e(E_0 + E_1) \tag{198}$$

where we have taken the pump $\mathbf{E}_0$ to be directed along the x-direction.

The electron motion has two parts: a high-frequency part, in which the electron move independently of the ions, and a low-frequency part, in which they move in concert with the ions in a quasineutral manner.

Again, one has for the high-frequency part of the electron motion, the oscillating equilibrium:

$$\frac{\partial V_{e_0}}{\partial t} = -\frac{e}{m_e} E_0 = -\frac{e}{m_e} \bar{E}_0 \cos \omega_0 t. \tag{199}$$

Linearizing about this equilibrium, one has for the motion induced by the space-charge field:

$$\frac{\partial V_{e_1}}{\partial t} = -\frac{e}{m_e} E_1 \tag{200}$$

$$ikE_1 = -4\pi e n_{e_1} \tag{201}$$

from which,

$$\frac{\partial V_{e_1}}{\partial t} = \frac{4\pi e^2}{ikm_e} n_{e_1}. \tag{202}$$

Further, one has for the electrons:

$$\frac{\partial n_{e_1}}{\partial t} + ikV_{e_0} n_{e_1} + N_0 ikV_{e_1} = 0. \tag{203}$$

Splitting n_e again into a high-frequency part and a low-frequency part, i.e.,

$$n_{e_1} = n_{e_h} + n_{e_\ell}$$

and noting that $n_{e_\ell} \approx n_{i_1}$, equation (203) gives

$$\frac{\partial n_{e_h}}{\partial t} + ikV_{e_0} n_{i_1} + ikN_0 V_{e_1} = 0. \tag{204}$$

Using equations (199)—(202), equation (204) gives

$$\frac{\partial^2 n_{e_h}}{\partial t^2} + \omega_{p_e}^2 n_{e_h} = \frac{ike}{m_e} n_{i_1} E_0. \tag{205}$$

If $n_{e_h} \sim e^{-i\omega_0 t}$, one obtains from equation (205) and (201),

$$E_1 = -\frac{4\pi e^2}{m_e} \frac{n_{i_1} E_0}{\omega_{p_e}^2 - \omega_0^2}. \tag{206}$$

Using (196) and (206), equation (197) becomes

$$\frac{\partial^2 n_{i_1}}{\partial t^2} \approx \left(\frac{e^2 k^2}{2m_e m_i} \frac{\bar{E}_0^2}{\omega_{p_e}^2 - \omega_0^2} \right) n_{i_1}. \tag{207}$$

If

$$n_{i_1} \sim e^{\gamma t} \tag{208}$$

then equation (207) gives

$$\gamma^2 = \frac{e^2 k^2}{2m_e m_i} \frac{\bar{E}_0^2}{\omega_{p_e}^2 - \omega_0^2}. \tag{209}$$

Note that γ is real if $\omega_0 < \omega_{p_e}$, (so that the decay instability cannot then occur).

The effect of a weak ambient magnetic field on the oscillating two-stream and parametric decay instabilities was considered by Freund and Papadopoulos [153]. The latter found that the presence of an ambient magnetic field can substantially enhance thresholds and reduce growth rates for waves propagating at an oblique angle with respect to the ambient magnetic field. The oscillating two-stream instability of whistler waves was considered by Forslund *et al.* [154], and Lee [155], and that of ion-cyclotron waves by Tripathi and Liu [156].

Kim *et al.* [68] gave observations of the parametric instability in a driven nonuniform plasma with a pump electric field directed parallel to the density gradient which drives electrostatic waves interacting nonlinearly with the pump field. The ponderomotive force of the linearly enhanced electric field first generated a density cavity at the resonant location $\omega_{p_e} = \omega_0$. The cavity in turn trapped the rf field and caused mutual enhancements between the localized rf field and the density-perturbation which is a characteristic of the oscillating two-stream instability. The time — and space — resolved measurements showed simultaneous exponential growth of the electric field intensity $\langle E_T^2 \rangle$ and the density perturbation δn at the resonant location (Figure 4.12a), with the growth rate γ depending linearly on the pump power E_0^2 (Figure 4.12b). Rumsby and Michaelis [157] presented some results showing how the density profile of a laser-produced carbon plasma is modified when the plasma is irradiated with an intense pulse of radiation from a CO_2 laser. Initially the CO_2 radiation was absorbed by inverse bremsstrahlung in a region about 2 mm in front of the target surface where the electron density was about 10^{18} cm^{-3} and the electron temperature a few eV, (the absorption length ~0.5 mm). This region was rapidly heated and hence expanded radically and axially leading to a density profile which caused the beam to be trapped (see Figure 4.13). The beam then penetrated further in towards the target surface causing more heating until eventually it reached the resonant laser where $n_e \sim 10^{19}$ cm^{-3}. Here it was either reflected or strongly absorbed if a parametric instability was excited. Donaldson and Spalding [158] made measurements of electron density and X-ray emission on a plasma generated by a CO_2 laser beam, and observed a density cavity and X-ray filamentation near the resonant layer. In Figure 4.14a

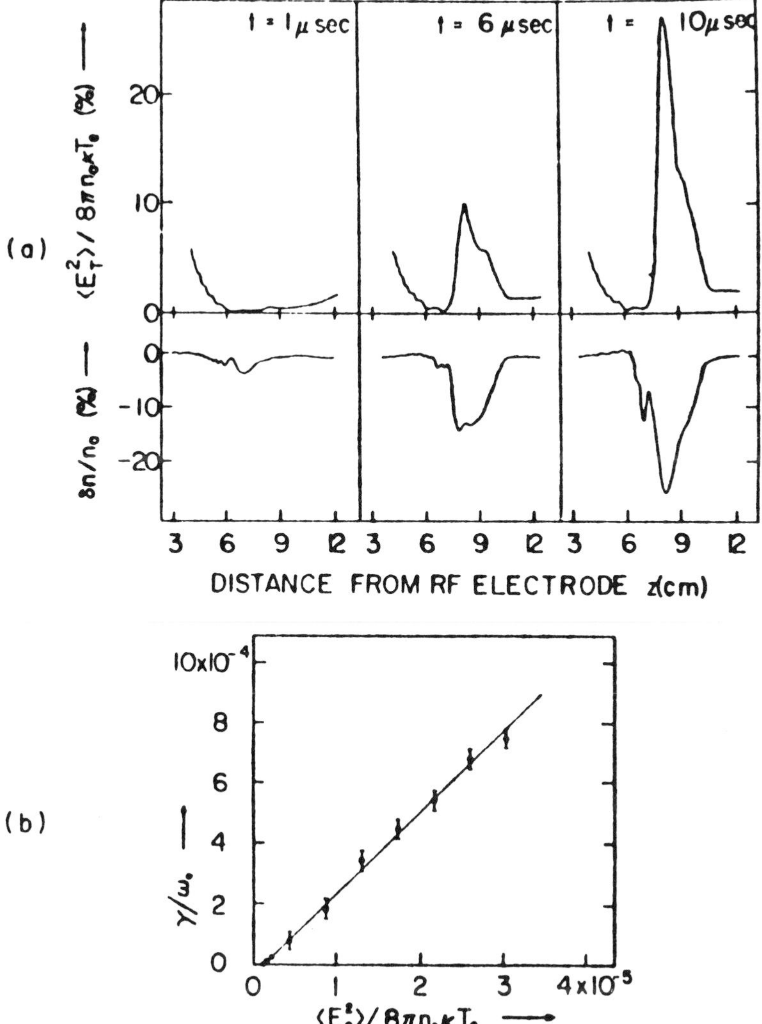

Figure 4.12. (a) Profiles of the normalized field intensity (top traces) and density perturbation (bottom traces) at different times t after turn-on of 10-W rf pump (measured with probe). rf pressure and density perturbation mutually enhance each other. (b) Growth rate of $\langle E_T^2 \rangle$ versus applied pump intensity $\langle E_0^2 \rangle$. (Due to Kim *et al.* [68], by courtesy of The American Physical Society.)

the density cavitation and in Figure 4.14b filamentation of the X-ray intensity are apparent.

(g) *The Question of the Time-Dependent Ponderomotive Force*

The formulation of the ponderomotive force when the electromagnetic field

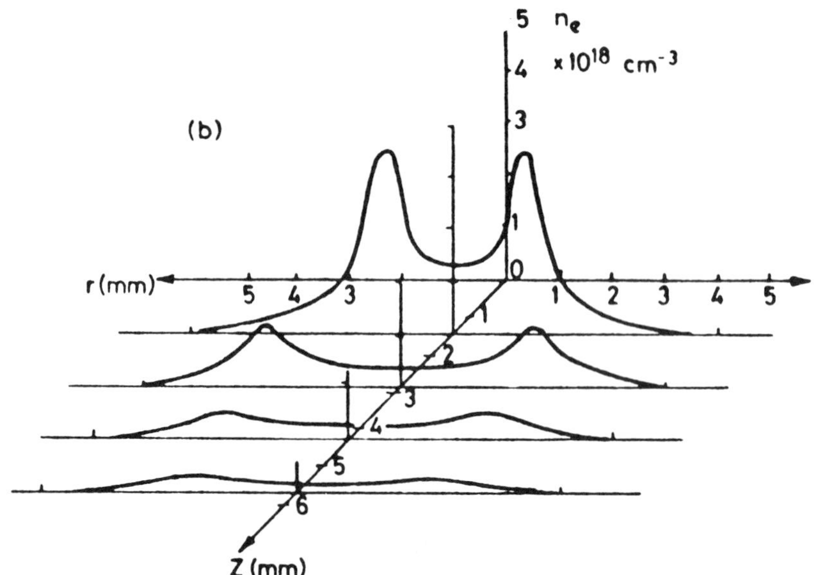

Figure 4.13. Density profiles. (Due to Rumsby and Michaelis [157], by courtesy of The American Physical Society.)

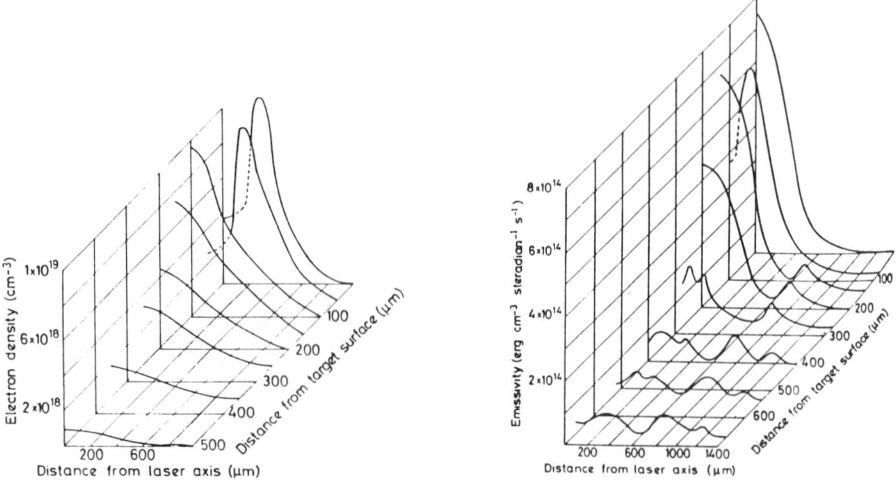

Figure 4.14. (a) Radial density profiles at $t = 25$ nsec. (b) Time-integrated radial x-ray emissivity profiles, after transmission through 0.564 mg cm^{-2} of Al filter. (Due to Donaldson and Spalding [158], by courtesy of The American Physical Society.)

amplitude has a slow time dependence was a controversial topic which has only recently been resolved (Kentwell and Jones [159]).

The equation of momentum balance for a plasma is

$$\frac{\partial}{\partial t}(n_\alpha \mathbf{V}_\alpha) + \nabla \cdot (n_\alpha \mathbf{V}_\alpha \mathbf{V}_\alpha) + \frac{e_\alpha n_\alpha}{m_\alpha}\left(\mathbf{E} + \frac{1}{c}\mathbf{V} \times \mathbf{B}\right) = 0. \tag{210}$$

Divide the various quantities into a fast-varying part and a slow-varying part:

$$q = \langle q \rangle + \tilde{q}. \tag{211}$$

The equation for the slow-varying part is then given by averaging over fast oscillations:

$$\frac{\partial}{\partial t}(\langle n_\alpha \rangle \langle \mathbf{V}_\alpha \rangle + \langle \tilde{n}_\alpha \tilde{\mathbf{V}}_\alpha \rangle) = \frac{e_\alpha \langle n_\alpha \rangle}{m_\alpha}\langle \mathbf{E} \rangle + \mathbf{f}_\alpha \tag{212}$$

where $\mathbf{f}_\alpha$ is the ponderomotive force given by

$$\mathbf{f}_\alpha = -\nabla \cdot \langle (\langle n_\alpha \rangle \tilde{\mathbf{V}}_\alpha \tilde{\mathbf{V}}_\alpha) \rangle + \frac{e_\alpha}{m_\alpha}\left(\langle \tilde{\mathbf{n}}_\alpha \tilde{\mathbf{E}} \rangle + \frac{1}{c}\langle n_\alpha \rangle \langle \tilde{\mathbf{V}}_\alpha \times \tilde{\mathbf{B}} \rangle\right). \tag{213}$$

Now, the zeroth approximation of equation (210) gives on Fourier transformation:

$$\tilde{\mathbf{V}}_\alpha(\omega) = \frac{i e_\alpha \tilde{\mathbf{E}}(\omega)}{m_\alpha \omega}. \tag{214}$$

Thus,

$$\begin{aligned} \nabla \cdot \langle n_0 \tilde{\mathbf{V}}_\alpha \tilde{\mathbf{V}}_\alpha \rangle &= \nabla \cdot \left\langle n_0 \int \frac{d\omega_1}{2\pi} \int \frac{d\omega_2}{2\pi} \tilde{\mathbf{V}}_\alpha(\omega_1) \tilde{\mathbf{V}}_\alpha(\omega_2) e^{-i(\omega_1 + \omega_2)t} \right\rangle \\ &= -\frac{1}{4\pi m_\alpha} \nabla \cdot \left\langle \int \frac{d\omega_1}{2\pi} \int \frac{d\omega_2}{2\pi} \frac{\omega_{p_\alpha}^2}{\omega_1 \omega_2} \times \right. \\ &\qquad \left. \times \tilde{\mathbf{E}}(\omega_1) \tilde{\mathbf{E}}(\omega_2) e^{-i(\omega_1 + \omega_2)t} \right\rangle. \end{aligned} \tag{215}$$

On putting

$$\omega_2 = -\omega_1 + \Omega, \quad \Omega \ll \omega_p \tag{216}$$

and letting

$$\varepsilon(\omega) = 1 - \frac{\omega_{p_\alpha}^2}{\omega^2} \tag{217}$$

equation (215) becomes

$$\nabla \cdot \langle n_0 \tilde{\mathbf{V}}_\alpha \tilde{\mathbf{V}}_\alpha \rangle = \frac{1}{4\pi m_\alpha} \nabla \cdot \left\langle \int \frac{d\omega_1}{2\pi} \int \frac{d\Omega}{2\pi} \cdot \frac{\omega_{p_\alpha}^2}{\omega_1^2} \left(1 + \frac{\Omega}{\omega_1}\right) \times \right.$$

$$\left. \times \tilde{\mathbf{E}}(\omega_1)\tilde{\mathbf{E}}(-\omega_1 + \Omega) e^{-i\Omega t} \right\rangle$$

$$= \frac{1}{4\pi m_\alpha} \nabla \cdot \left\langle \int \frac{d\omega_1}{2\pi} \int \frac{d\Omega}{2\pi} \left[1 - \varepsilon(\omega_1) \right.\right.$$

$$\left.\left. + \frac{\Omega}{2} \frac{\partial \varepsilon(\omega_1)}{\partial \omega_1} \right] \tilde{\mathbf{E}}(\omega_1)\tilde{\mathbf{E}}(-\omega_1 + \Omega) e^{-i\Omega t} \right\rangle$$

$$= \frac{1}{4\pi m_\alpha} \nabla \cdot \langle \tilde{\mathbf{E}}\tilde{\mathbf{E}} \rangle$$

$$- \frac{1}{8\pi m_\alpha} \nabla \cdot \left\langle \int \frac{d\omega_1}{2\pi} \int \frac{d\Omega}{2\pi} \left[\varepsilon(\omega_1) \right.\right.$$

$$\left.\left. + \varepsilon(\omega_1 - \Omega) \right] \tilde{\mathbf{E}}(\omega_1)\tilde{\mathbf{E}}(-\omega_1 + \Omega) e^{-i\Omega t} \right\rangle$$

$$= \frac{1}{4\pi m_\alpha} \nabla \cdot \langle \tilde{\mathbf{E}}\tilde{\mathbf{E}} \rangle - \frac{1}{8\pi m_\alpha} \nabla \cdot \langle \tilde{\mathbf{D}}\tilde{\mathbf{E}} + \tilde{\mathbf{E}}\tilde{\mathbf{D}} \rangle. \tag{218}$$

Further,

$$\frac{e_\alpha}{m_\alpha c} \langle n_0 \tilde{\mathbf{V}}_\alpha \times \tilde{\mathbf{B}} \rangle = -\frac{1}{4\pi} \left\langle \nabla \cdot (\tilde{\mathbf{B}}\tilde{\mathbf{B}}) - \frac{1}{2} \nabla (\tilde{\mathbf{B}} \cdot \tilde{\mathbf{B}}) \right.$$

$$\left. - \frac{1}{c} \frac{\partial \tilde{\mathbf{E}}}{\partial t} \times \tilde{\mathbf{B}} \right\rangle \tag{219}$$

$$e_\alpha \langle \tilde{\mathbf{n}}_\alpha \tilde{\mathbf{E}} \rangle = -\frac{1}{4\pi} \langle (\nabla \cdot \tilde{\mathbf{E}})\tilde{\mathbf{E}} \rangle$$

$$= -\frac{1}{4\pi} \left\langle \nabla \cdot (\tilde{\mathbf{E}}\tilde{\mathbf{E}}) - \frac{1}{2} \nabla (\tilde{\mathbf{E}} \cdot \tilde{\mathbf{E}}) + \tilde{\mathbf{E}} \times (\nabla \times \tilde{\mathbf{E}}) \right\rangle \tag{220}$$

where we have used Maxwell's equations:

$$\nabla \times \tilde{\mathbf{B}} = \frac{1}{c}\frac{\partial \tilde{\mathbf{E}}}{\partial t} + \frac{4\pi e_\alpha}{c}\langle n_\alpha \rangle \tilde{\mathbf{V}}_\alpha \tag{221}$$

$$\nabla \cdot \tilde{\mathbf{E}} = 4\pi e_\alpha \tilde{n}_\alpha. \tag{222}$$

Using (218)—(220), and the other Maxwell equation:

$$\nabla \times \tilde{\mathbf{E}} = -\frac{1}{c}\frac{\partial \tilde{\mathbf{B}}}{\partial t} \tag{223}$$

(213) becomes

$$\begin{aligned}
\mathbf{f}_\alpha = {} & \frac{1}{4\pi}\nabla \cdot \left\langle \tilde{\mathbf{B}}\tilde{\mathbf{B}} + \frac{1}{2}(\tilde{\mathbf{E}}\tilde{\mathbf{D}} + \tilde{\mathbf{D}}\tilde{\mathbf{E}}) - \frac{1}{2}(\tilde{\mathbf{E}}\cdot\tilde{\mathbf{E}} + \tilde{\mathbf{B}}\cdot\tilde{\mathbf{B}})\mathbf{I}\right\rangle \\
& - \frac{1}{4\pi c}\frac{\partial}{\partial t}(\tilde{\mathbf{E}}\times\tilde{\mathbf{B}}) \\
= {} & -\frac{\omega_{p_\alpha}^2}{16\pi\omega^2}\nabla|\tilde{\mathbf{E}}|^2 + \frac{1}{4\pi c}\frac{\partial}{\partial t}\langle \tilde{\mathbf{D}}\times\tilde{\mathbf{B}} - \tilde{\mathbf{E}}\times\tilde{\mathbf{B}}\rangle \\
& - \frac{1}{8\pi}\langle\nabla\times(\tilde{\mathbf{E}}\times\tilde{\mathbf{D}})\rangle \\
= {} & -\frac{\omega_{p_\alpha}^2}{16\pi\omega^2}\nabla|\tilde{\mathbf{E}}|^2 + \frac{i\omega_{p_\alpha}^2}{16\pi\omega^3}\left[\nabla\left(\tilde{\mathbf{E}}^*\cdot\frac{\partial\tilde{\mathbf{E}}}{\partial t} - \tilde{\mathbf{E}}\cdot\frac{\partial\tilde{\mathbf{E}}^*}{\partial t}\right)\right. \\
& \left. + \frac{\partial}{\partial t}\{(\tilde{\mathbf{E}}\cdot\nabla)\tilde{\mathbf{E}}^* - (\tilde{\mathbf{E}}^*\cdot\nabla)\tilde{\mathbf{E}}\} + \frac{\partial}{\partial t}\nabla\times(\tilde{\mathbf{E}}\times\tilde{\mathbf{E}}^*)\right] \\
= {} & -\frac{\omega_{p_\alpha}^2}{16\pi\omega^2}\nabla|\mathbf{E}|^2 + \frac{i\omega_{p_\alpha}^2}{16\pi\omega^3}\left[\frac{\partial}{\partial t}\{\tilde{\mathbf{E}}(\nabla\cdot\tilde{\mathbf{E}}^*)\right. \\
& \left. - \tilde{\mathbf{E}}^*(\nabla\cdot\tilde{\mathbf{E}})\} + \nabla\left(\tilde{\mathbf{E}}^*\cdot\frac{\partial\tilde{\mathbf{E}}}{\partial t} - \tilde{\mathbf{E}}\cdot\frac{\partial\tilde{\mathbf{E}}^*}{\partial t}\right)\right]
\end{aligned} \tag{224}$$

where we have used the relation

$$\tilde{\mathbf{D}} = \left(\varepsilon\tilde{\mathbf{E}} + i\frac{\partial\varepsilon}{\partial\omega}\frac{\partial\tilde{\mathbf{E}}}{\partial t}\right)e^{-i\omega t} + \text{c.c.} \tag{225}$$

It may be noted that because of the earlier disagreement in deciding which terms should be included in the definition of the ponderomotive force as well as some erroneous developments, different results were given by Kono *et al.* [145], Tskhakaya [148], Karpman and Shagalov [150].

Epilogue to Parametric Processes. In a plasma acted on by an intense laser field, one of the products is a spectrum of excited Langmuir waves. As the amplitudes of these excited Langmuir waves build up they too can undergo further decay instabilities and induced scattering on the particles with the consequent excitation of Langmuir wave of smaller wavenumbers and lower frequencies (and of course an ion-acoustic wave to satisfy the matching conditions). For a hot, tenuous plasma collisions between the particles will be negligible, and as the wavenumber of the Langmuir waves decreases the effect of Landau damping will also become negligible. A problem then arises as to how this large amount of energy concentrated in the low-wave-number end of the Langmuir spectrum called Langmuir wave condensate will be dissipated. Zakharov [44, 45] proposed that this spectral concentration of energy would induce a nonlinear modulational instability that leads to the formation of spatialized localized-field structures of shorter scales which then collapse and the wave energy is dissipated into particle energy. Such a growth and decay of these localized field structures would provide an efficient mechanism of transferring energy from small wavenumbers to large wavenumbers, where it would be dissipated by Landau damping. So, we will now take up the problem of modulational instability of a large-amplitude plasma wave and the formation of envelope solitons and their eventual collapse and dissipation.

V

Modulational Instability and Envelope Solitons

Let us superimpose a slowly-varying modulation on a stationary wave, and study the evolution of such a modulation. Let us assume that the wave can still be taken to be sinusoidal, i.e.,

$$\phi = a_0 \cos \theta_0, \quad \theta_0 = k_0 x - \omega_0 t. \tag{1}$$

Because of the nonlinearities, the dispersion relation is of the form:

$$\omega_0 = \omega_0(k_0, a_0^2). \tag{2}$$

Consequent to the superimposition of the modulation, let us assume that the wave is still plane periodic, but with the amplitude and phase varying slowly in x and t:

$$a = a(x, t), \theta(x, t) = k_0 x - \omega_0 t + \varphi(x, t). \tag{3}$$

Thus, one may define generalized frequency and wavenumber:

$$\begin{aligned} \omega(x, t) &= -\theta_t = \omega_0 - \varphi_t \\ k(x, t) &= \theta_x = k_0 + \varphi_x. \end{aligned} \tag{4}$$

For weak modulations, one may write

$$\omega = \omega_0 + \frac{\partial \omega}{\partial a_0^2} \cdot (a^2 - a_0^2) + \frac{\partial \omega}{\partial k_0} \cdot (k - k_0)$$
$$+ \frac{\partial^2 \omega}{\partial k_0^2} \cdot \frac{1}{2} (k - k_0)^2 + \cdots \tag{5}$$

so that, using (4), one obtains

$$\varphi_t + u_0 \varphi_x + \frac{u_0'}{2} \varphi_x^2 + \frac{\partial \omega}{\partial a_0^2} \cdot (a^2 - a_0^2) = 0 \tag{6}$$

where u_0 is the group velocity,

$$u_0 \equiv \partial \omega / \partial k_0.$$

Next, noting from (4) that

$$u \equiv \frac{\partial \omega}{\partial k} = u_0 + u_0' \cdot \varphi_x \tag{7}$$

and using the well-known result (Whitham [160]),

$$\frac{\partial a^2}{\partial t} + \frac{\partial}{\partial x}(ua^2) = 0 \tag{8}$$

one obtains

$$\frac{\partial a^2}{\partial t} + u_0 \frac{\partial a^2}{\partial x} + u_0' \frac{\partial}{\partial x}(\varphi_x a^2) = 0. \tag{9}$$

Introduce,

$$\xi = x - u_0 t, \quad \tau = u_0' t \tag{10}$$

so that equations (6) and (9) become

$$\varphi_\tau + \frac{1}{2}\varphi_\xi^2 + \frac{1}{u_0'}\frac{\partial \omega}{\partial a_0^2} \cdot (a^2 - a_0^2) = 0 \tag{11}$$

$$(a^2)_\tau + (a^2 \varphi_\xi)_\xi = 0. \tag{12}$$

Let

$$\varphi, (a^2 - a_0^2) \sim e^{i(Kx - \Omega\tau)}. \tag{13}$$

One then obtains on linearizing equations (11) and (12),

$$\Omega^2 = 2\frac{a_0^2}{u_0'}\frac{\partial \omega}{\partial a_0^2} K^2. \tag{14}$$

Therefore, instability arises if

$$\frac{1}{u_0'}\frac{\partial \omega}{\partial a_0^2} < 0. \tag{15}$$

As an illustration of application of the above formula, let us consider the modulational stability of nonlinear electron plasma waves.

Modulational Stability of Nonlinear Electron Plasma Waves

Modulational stability of nonlinear waves in a warm electron plasma was investigated by Infeld and Rowlands [161] and Shivamoggi [162]. We will establish here the modulational stability of nonlinear waves in a warm electron plasma by explicitly evaluating the amplitude-dependent nonlinear dispersion relation for a stationary traveling wave.

Consider a one-dimensional wave motion in a warm isothermal electron plasma, with the ions forming an immobile neutralizing background. The equations governing such a wave motion are:

$$\frac{\partial n}{\partial t} + \frac{\partial}{\partial x}(nv) = 0 \tag{16}$$

$$\frac{\partial v}{\partial t} + v\frac{\partial v}{\partial x} = -\frac{1}{n}\frac{\partial n}{\partial x} + \frac{\partial \phi}{\partial x} \tag{17}$$

$$\frac{\partial^2 \phi}{\partial x^2} = n - 1 \tag{18}$$

where n is the electron number density normalized by the mean number density n_0, v is the electron fluid velocity non-dimensionalized by the electron thermal speed $V_{T_e} = \sqrt{KT_e/m_e}$, ϕ is the electric potential normalized by T_e/e, t is normalized by ω_p^{-1} and x by the Debye length $\lambda_D = V_{T_e}/\omega_p$. T_e is the temperature of the electron plasma and K is the Boltzmann constant.

If we look for a stationary wave traveling in the x-direction, the various physical quantities will depend on x and t only through the combination $\xi = kx - \omega t$. Then on putting

$$n = 1 + N \tag{19}$$

we may derive from equations (16)—(18):

$$\left[\frac{\omega^2}{(1+N)^3} - \frac{k^2}{1+N}\right]\frac{\partial^2 N}{\partial \xi^2} + \left[\frac{k^2}{(1+N)^2} - \frac{3\omega^2}{(1+N)^4}\right]\left(\frac{\partial N}{\partial \xi}\right)^2 + N = 0. \tag{20}$$

Let us now introduce a small parameter $\varepsilon \ll 1$ which may characterize a typical wave amplitude, and seek solutions to equation (20) of the form

$$N(\xi;\varepsilon) = \sum_{n=1}^{\infty} \varepsilon^n N_n(\xi)$$
$$\omega(k;\varepsilon) = \sum_{n=0}^{\infty} \varepsilon^n \omega_n(k). \tag{21}$$

We are using the method of strained parameters.

Using (21), equation (20) gives

$$(\omega_0^2 - k^2)\frac{\partial^2 N_1}{\partial \xi^2} + N_1 = 0 \tag{22}$$

$$(\omega_0^2 - k^2)\frac{\partial^2 N_2}{\partial \xi^2} + N_2 = (3\omega_0^2 - k^2)\left[N_1 \frac{\partial^2 N_1}{\partial \xi^2} + \left(\frac{\partial N_1}{\partial \xi}\right)^2\right] - 2\omega_0\omega_1 \frac{\partial^2 N_1}{\partial \xi^2} \tag{23}$$

$$\begin{aligned}(\omega_0^2 - k^2)\frac{\partial^2 N_3}{\partial \xi^2} + N_3 = {} & (3\omega_0^2 - k^2)\left[N_2 \frac{\partial^2 N_1}{\partial \xi^2} + N_1 \frac{\partial^2 N_2}{\partial \xi^2} + 2\frac{\partial N_1}{\partial \xi}\frac{\partial N_2}{\partial \xi}\right] \\ & - (6\omega_0^2 - k^2)\left[N_1^2 \frac{\partial^2 N_1}{\partial \xi^2} + 2N_1\left(\frac{\partial N_1}{\partial \xi}\right)^2\right] \\ & - (2\omega_0\omega_2 + \omega_1^2)\frac{\partial^2 N_1}{\partial \xi^2} - 2\omega_0\omega_1 \frac{\partial^2 N_2}{\partial \xi^2}.\end{aligned} \tag{24}$$

We obtain from equation (22), the linear result:

$$\begin{aligned} N_1 &= a_0 \cos \xi \\ \omega_0^2 &= 1 + k^2. \end{aligned} \tag{25}$$

Using (25), the removal of secular terms on the right-hand side of equation (23) requires

$$\omega_1 \equiv 0 \tag{26}$$

and then the solution to equation (23) is given by

$$N_2 = \frac{a_0^2}{3}(3 + 2k^2)\cos 2\xi. \tag{27}$$

Using (25) through (27), the condition for the removal of secular terms on the right-hand side of the equation (24) gives

$$\omega_2 = \frac{k^2(8k^2 + 9)}{24\sqrt{1 + k^2}}\, a_0^2. \tag{28}$$

Thus, the nonlinear dispersion relation is given by

$$\omega^2 = (1 + k^2) + \varepsilon^2 \frac{k^2}{12}(8k^2 + 9)a_0^2 + 0(\varepsilon^3). \tag{29}$$

Since, in this case

$$\frac{1}{u_0'}\,\frac{\partial \omega}{\partial a_0^2} = \frac{1}{12}(1 + k^2)^{3/2}(8k^2 + 9) > 0 \tag{30}$$

we conclude that the nonlinear electron plasma waves are modulationally stable.

Envelope Solitons

If in (5), we replace $(\omega - \omega_0)$ by the operator $i\,\partial/\partial t$ and $(k - k_0)$ by $-i\,\partial/\partial x$, we obtain the nonlinear Schrödinger equation (Karpman and Krushkal [163]).

$$i\left[\frac{\partial a}{\partial t} + \frac{\partial \omega}{\partial k_0}\frac{\partial a}{\partial x}\right] + \frac{1}{2}\frac{\partial^2 \omega}{\partial k_0^2}\frac{\partial^2 a}{\partial x^2} - \frac{\partial \omega}{\partial a_0^2}|a|^2 a = 0. \tag{31}$$

In the frame of reference moving with the group velocity, equation (31) becomes

$$i\frac{\partial a}{\partial \tau} + \frac{1}{2}\frac{\partial^2 a}{\partial \xi^2} + \chi|a|^2 a = 0 \tag{32}$$

where,

$$\chi \equiv -\frac{\partial \omega/\partial a_0^2}{\partial^2 \omega/\partial k_0^2}.$$

Plane-wave solutions of equation (17) are modulationally unstable if $\chi > 0$ i.e., a ripple on the envelope of the wave will tend to grow. Physically this nonlinear instability may be interpreted as follows. The gradient on wave intensity causes a ponderomotive force which moves both electrons and ions toward the intensity minima, forming a ripple in the plasma density. Plasma waves are trapped in regions of low density because their dispersion relation

$$\omega^2 = \omega_p^2 + 3k^2 V_T^2$$

allows waves of large k to exist only in regions of small ω_p. The trapping of part of the k-spectrum further enhances the wave intensity in the regions where it was already high, thus causing the envelope to develop a growing ripple.

In order to see why the sign of χ matters, consider a ripple on the envelope. Suppose $\partial\omega/\partial a_0^2 < 0$. Then the phase velocity $\omega/k = V_p$ becomes somewhat smaller in the region of high intensity. This causes the wave crests to pile up on the left in Figure (5.1) and to spread out on the right. If $\partial^2\omega/\partial k^2 > 0$, the group velocity $u = \partial\omega/\partial k$ will be larger on the left than on the right, so the wave energy will pile up into a smaller space. Thus, the ripple on the envelope will become narrower and larger as shown below. If $\partial\omega/\partial a_0^2$ and $\partial^2\omega/\partial k^2$ were of the same sign, this modulational instability will not develop. Alternatively, when $\chi > 0$, one may regard equation (17) as the Schrödinger equation for 'quasi-particles' whose wave function is given by a, and which are trapped by a self-generated localized potential well $V = \chi|a|^2$. If $\chi > 0$, this potential has an attractive sign. If the 'quasi-particle' density $|a|^2$ increases, the potential depth increases and if $\chi > 0$, more 'quasi-particles' are attracted, leading to a further increase in the potential depth. In this sense, the instability may be regarded as a consequence of the self-trapping of the 'quasi-particles'.

Let us now construct a localized stationary solution of the nonlinear

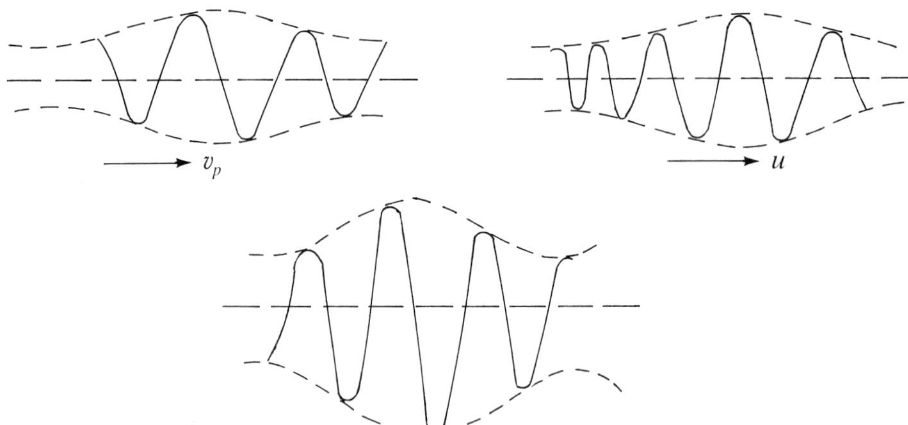

Figure 5.1. Modulational instability. (Due to Chen [141], by courtesy of Plenum Press.)

Schrödinger equation (32). Put

$$a = v(\xi - U\tau)e^{i(\gamma\xi - s\tau)} \tag{33}$$

so that equation (32) gives, (primes denote differentiation with respect to argument),

$$\frac{1}{2}v'' + \frac{i}{2}(2\gamma - U)v' + \left(s - \frac{\gamma^2}{2}\right)v + \chi|v|^2v = 0. \tag{34}$$

Let,

$$\gamma = \frac{U}{2}, \quad s = \frac{U^2}{8} - \frac{\alpha}{2} \tag{35}$$

so that equation (34) becomes

$$v'' - \alpha v + 2\chi v^3 = 0. \tag{36}$$

Upon integrating once, equation (36) gives

$$v'^2 = \alpha v^2 - \chi v^4 \tag{37}$$

from which,

$$\int (\alpha - \chi v^2)^{-1/2} \frac{\mathrm{d}v}{v} = \xi - U\tau. \tag{38}$$

Put,

$$w = \left(1 - \frac{\chi v^2}{\alpha}\right)^{1/2} \tag{39}$$

then (38) gives

$$\frac{1}{2}\ln\frac{1-w}{1+w}=\sqrt{\alpha}\,(\xi-U\tau). \tag{40}$$

If $\chi > 0, \alpha > 0$, one obtains from (40)

$$v=\sqrt{\frac{\alpha}{\chi}}\,\operatorname{sech}\sqrt{\alpha}\,(\xi-U\tau) \tag{41}$$

which represents an envelope soliton (Figure 5.2) that propagates unchanged in shape and with constant velocity. The latter result arises because the non-linearity and dispersion exactly balance each other — a result which turns out to be unique to one-dimensional solutions. If the wave energy is to move along with the above envelope, U must be near the group velocity of the wave in question.

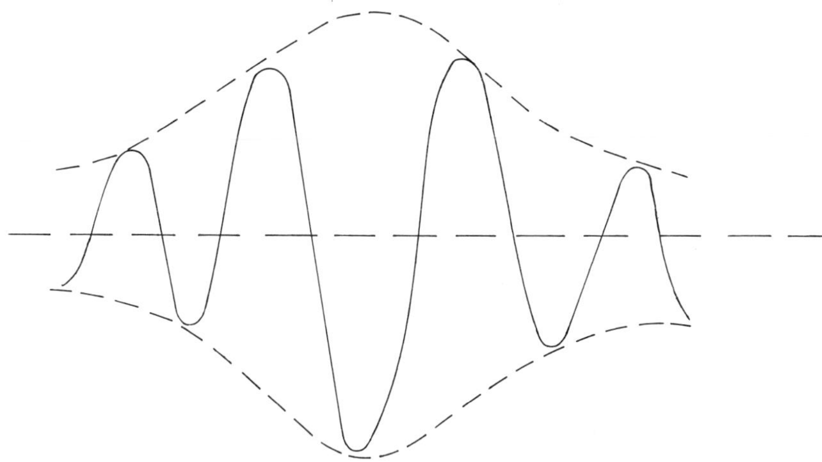

Figure 5.2. An envelope soliton.

Note that this solution is possible only if modulational instability occurs, i.e., if $\chi > 0$. This suggests that the end result of an instable wavetrain subject to small modulations is a series of envelope solitons.

Using the inverse-scattering formalism, Zakharov and Shabat [54] solved the nonlinear Schrödinger equation exactly for initial conditions which approach zero rapidly as $|x| \to \infty$. Zakharov and Shabat [54] found that:

(i) an initial wave envelope pulse of arbitrary shape will eventually disintegrate into a number of solitons of shorter scales and an oscillatory tail; (the number and structure of these solitons and the structure of the tail are completely determined by the initial conditions);
(ii) the tail is relatively small and disperses linearly resulting in $1/\sqrt{t}$ amplitude decay;

(iii) each soliton is a permanent progressive wave solution:
(iv) the solitons are stable in the sense that they can survive interactions with each other with no permanent change except a possible shift in position and phase.

Another property exhibited by the solutions of the nonlinear Schrödinger equation is that of recurrence of states. The long-time one-dimensional evolution of an initially uniform envelope was studied numerically by Morales *et al.* [55]. The growing sinusoidal standing-wave perturbation steepened with time into a shape resembling an elliptic function, but this structure then gave way back to a state resembling the original uniform envelope. The system continued to show a recurrent behavior. The latter arises from the fact that many nonlinear wave systems behave as if their 'effective' number of degrees of freedom is limited. This finite set of modes interact nonlinearly perpetually exchanging energy with each other. In conservative systems, the existence of invariants (Gibbons *et al.* [58]) plays a vital role in establishing recurrence properties (Thyagaraja [56, 57]). In order to demonstrate this, since recurrent behavior is typical only for bounded or periodic domains, let us restrict attention to the consideration of the initial-value problem with periodic boundary conditions:

$$\begin{aligned} i\,\frac{\partial\psi}{\partial t} &= \frac{\partial^2\psi}{\partial x^2} + \mu|\psi|^2\psi \\ x = 0, 1:\quad \psi, &\ \frac{\partial\psi}{\partial x} = 0. \end{aligned} \tag{42}$$

Let us assume that this initial-boundary-value problem has smooth and unique solutions for $t > 0$. In order to examine the qualitative properties of the solution, Thyagaraja [56, 57] derived certain elementary *a priori* bounds involving the integral invariants associated with this equation.

Two invariants with this problem are,

$$I(t) = \int_0^1 |\psi(x,t)|^2\,\mathrm{d}x \tag{43}$$

$$J(t) = \int_0^1 \left[|\psi_x(x,t)|^2 - \frac{\mu}{2}|\psi(x,t)|^4 \right] \mathrm{d}x \tag{44}$$

(there are actually an infinite number of integral invariants, as shown by Zakharov and Shabat [54]). Let us now consider any function $\psi(x, t)$ (not necessarily a solution of this equation) which is defined for $t > 0$ and is sufficiently smooth, and satisfies the boundary conditions and evolves in such a manner that the functionals $I(t)$ and $J(t)$ are constant. Consider the Rayleigh quotient defined by

$$Q(t)I(t) = \int_0^1 |\psi_x(x,t)|^2\,\mathrm{d}x. \tag{45}$$

Let,

$$t = 0: \qquad I = I_0, \quad J = J_0. \tag{46}$$

When $\mu < 0$, one has

$$Q(t) < \frac{J_0}{I_0}. \tag{47}$$

In order to treat cases with $\mu > 0$, at any instant t, let x_0 be the point at which $|\psi(x, t)|^2$ takes its minimum value. One then has

$$[\psi(x, t)]^2 = [\psi(x_0, t)]^2 + 2\int_{x_0}^{x} \psi\psi_x \, \mathrm{d}x \tag{48}$$

from which,

$$|\psi(x, t)|^2 \leqslant |\psi(x_0, t)|^2 + 2\int_0^1 |\psi\psi_x| \, \mathrm{d}x \tag{49}$$

which in turn gives

$$|\psi(x, t)|^2 \leqslant I_0 + 2I_0[Q(t)]^{1/2} \tag{50}$$

where we have used the result

$$\int_0^1 |\psi\psi_x| \, \mathrm{d}x \leqslant I(t)\,[Q(t)]^{1/2}. \tag{51}$$

On multiplying (50) through by $|\psi(x, t)|^2$ and integrating over x, one obtains

$$\int_0^1 |\psi(x, t)|^4 \, \mathrm{d}x \leqslant I_0^2 + 2I_0^2 Q^{1/2}. \tag{52}$$

Thus,

$$Q(t) \leqslant \frac{J_0}{I_0} + \frac{\mu}{2} I_0 + \mu I_0[Q(t)]^{1/2} \tag{53}$$

from which,

$$Q(t) \leqslant [M(I_0, J_0, \mu)]^2 \tag{54}$$

where,

$$M^2 - \mu I_0 M - \left(\frac{J_0}{I_0} + \frac{\mu}{2} I_0\right) = 0. \tag{55}$$

Using (54), (50) gives

$$\|\psi\|_\infty^2 \equiv \max_{0 \leqslant x \leqslant 1} |\psi(x, t)|^2 \leqslant I_0(1 + 2M). \tag{56}$$

Now, a solution $\psi(x, t)$ of equation (42) is said to be Lagrange stable if there exists a constant K independent of t such that

$$\|\psi\|_{\infty}^{2} \leqslant K. \tag{57}$$

Thus, (57) sets forth the Lagrange stable nature of the solutions of the one-dimensional nonlinear Schrödinger equation.

In order to interpret (54), let us expand $\psi(x, t)$ in a Fourier series in x,

$$\psi(x, t) = \sum_{n=-\infty}^{\infty} c_n(t) e^{2n\pi i x} \tag{58}$$

so that

$$I_0 = \sum_{n=-\infty}^{\infty} |c_n(t)|^2 \tag{59}$$

and

$$Q(t) = \frac{4\pi^2 \sum_{n=-\infty}^{\infty} n^2 |c_n(t)|^2}{\sum_{n=-\infty}^{\infty} |c_n(t)|^2}. \tag{60}$$

In analogy with quantum mechanics, one may interpret $Q(t)/4\pi^2$ as the instantaneous average of n^2. $M/2\pi$ can then be interpreted as an upper bound to the rms value of n, the number of modes carrying the 'wave energy' I_0. If we call $Q/2\pi$, the number of 'effective modes' and denote it by N_{ef}, (54) says that N_{ef} is bounded by a number of $M/2\pi$, where $M = M(I_0, J_0, \mu)$.

The modification of the nonlinear Schrödinger equation due to damping and its consequences were considered by Pereira and Stenflo [164], and Pereira [165].

(i) Langmuir Solitons

Modulational instability of Langmuir waves arises through local depressions in density and corresponding increases in the energy density of the Langmuir electric field (Zakharov [45]). The density depressions are produced by the low-frequency ponderomotive force due to the beating of the Langmuir waves with each other on electrons, expelling them (and hence the ions as well through ambipolar effects) from the region of strong fields. Such density cavities act as potential wells and trap the Langmuir electric field and intensify them. As the electric field grows, more plasma is removed and the field is trapped and intensified further until a collapse of the field may occur in two and three-dimensional cases — this result has been verified by the numerical solutions of an initial-value problem associated with Zakharov's equations (Pereira *et al.*

[47]) Nicholson *et al.* [48], Hafizi *et al.* [49]). In actuality, other nonlinearities may limit this field growth. Besides, dissipation due to Landau damping begins to cut in at short spatial scales — on the order of the Debye length, which can cause the collapsing wave packet to burn out in a finite time. In a one-dimensional case, however, it is possible that the rate at which the field energy is trapped within the cavity can be balanced by the rate at which the field energy can leak out due to convection. Under such circumstances, a spatially localized stationary field structure can result.

Mathematically, the Langmuir solitons are localized stationary solutions of the nonlinear Schrödinger equation for the Langmuir field with an effective potential proportional to the low-frequency electron-density perturbation. The latter is in turn governed by an equation for ion-acoustic waves driven by the ponderomotive force associated with the Langmuir field.

Consider one-dimensional Langmuir waves travelling in a plasma consisting of cold ions and hot electrons. One has for the electrons

$$\frac{\partial n_e}{\partial t} + \frac{\partial}{\partial x}(n_e V_e) = 0 \tag{61}$$

$$\frac{\partial V_e}{\partial t} + V_e \frac{\partial V_e}{\partial x} = -\frac{e}{m_e} E - \frac{3V_{T_e}^2}{n_e} \frac{\partial n_e}{\partial x} \tag{62}$$

$$\frac{\partial E}{\partial x} = 4\pi e(n_i - n_e). \tag{63}$$

We have assumed that the electrons respond adiabatically to high-frequency motions.

Let us decompose the various quantities into a high-frequency part and a low-frequency part characterized by time scales $\omega_{p_e}^{-1}$, $\omega_{p_i}^{-1}$, respectively:

$$E = E_h + E_\ell, \quad n_e = n_0 + n_\ell + n_{e_h}, \quad n_i = n_0 + n_\ell. \tag{64}$$

Where n_ℓ, E_ℓ refer to the low-frequency perturbation, and n_{e_h}, E_h refer to the high-frequency perturbation. Note that we have assumed charge neutrality in the low-frequency motions. Equations (61) and (63) then give

$$\frac{\partial n_{e_h}}{\partial t} + \frac{\partial}{\partial x}[(n_0 + n_\ell)V_e] = 0 \tag{65}$$

$$\frac{\partial V_e}{\partial t} = -\frac{e}{m_e} E_h - \frac{3V_{T_e}^2}{(n_0 + n_\ell)} \frac{\partial n_{e_h}}{\partial x} \tag{66}$$

$$\frac{\partial E_h}{\partial x} = -4\pi e n_{e_h}. \tag{67}$$

One obtains from equations (65) and (66):

$$\frac{\partial^2}{\partial t^2} n_{e_h} - (n_0 + n_\ell)\left[\frac{e}{m_e}\frac{\partial E_h}{\partial x} + \frac{3V_{T_e}^2}{(n_0 + n_\ell)}\frac{\partial^2}{\partial x^2} n_{e_h}\right] = 0. \tag{68}$$

Using equation (67), equation (68) gives

$$\frac{\partial^2 E_h}{\partial t^2} + \omega_{p_e}^2 \left(1 + \frac{n_\ell}{n_0}\right) E_h - 3V_{T_e}^2 \frac{\partial^2 E_h}{\partial x^2} = 0. \tag{69}$$

Equation (69) describes the trapping of a Langmuir wave packet in density cavities.

Let us look for a solution of the form

$$E_h(x, t) = \mathscr{E}(x, t) e^{-i\omega_{p_e} t} \tag{70}$$

where $\mathscr{E}(x, t)$ has a slow-variation in time. Equation (69) then gives

$$i\omega_{p_e} \frac{\partial \mathscr{E}}{\partial t} + \frac{3}{2} V_{T_e}^2 \frac{\partial^2 \mathscr{E}}{\partial x^2} = \frac{1}{2} \omega_{p_e}^2 \frac{n_\ell}{n_0} \mathscr{E}. \tag{71}$$

Equation (71) is based on the premise that the effects of nonlinearity and finite temperature enter as corrections of same order to the linear cold-plasma dispersion relation near cutoff values of the wavenumber.

The low-frequency motions are governed by

$$\frac{e}{m_e} E_\ell = -\frac{V_{T_e}^2}{n_0 + n_\ell} \frac{\partial n_\ell}{\partial x} - V_e \frac{\partial V_e}{\partial x} \tag{72}$$

$$\frac{\partial n_\ell}{\partial t} + \frac{\partial}{\partial x} [(n_0 + n_\ell) V_i] = 0 \tag{73}$$

$$\frac{\partial V_i}{\partial t} = \frac{e}{m_i} E_\ell \tag{74}$$

where we have neglected to electron inertia and have assumed the electrons to respond isothermally to low-frequency motions.

One obtains from equations (73) and (74):

$$\frac{\partial^2 n_\ell}{\partial t^2} + \frac{\partial}{\partial x} \left[(n_0 + n_\ell) \frac{e}{m_e} E_\ell \right] = 0. \tag{75}$$

Using equation (72), equation (75) becomes

$$\frac{\partial^2 n_\ell}{\partial t^2} - C_s^2 \frac{\partial^2 n_\ell}{\partial x^2} = \frac{n_0 m_e}{m_i} \frac{\partial}{\partial x} \left(V_e \frac{\partial V_e}{\partial x} \right). \tag{76}$$

The low-frequency contribution of the term $V_e(\partial V_e/\partial x)$ is obtained by averaging it over the time scale ω_{pe}^{-1}:

$$\begin{aligned} \left\langle V_e \frac{\partial V_e}{\partial x} \right\rangle &= \frac{1}{4} \frac{\partial}{\partial x} \langle | V_e|^2 \rangle \\ &= \frac{e^2}{4\omega_{p_e}^2 m_e^2} \frac{\partial}{\partial x} (| E_h|^2) \\ &= \frac{1}{16\pi m_e n_0} \frac{\partial}{\partial x} (|\mathscr{E}|^2). \end{aligned} \tag{77}$$

Equation (77) represents the fast-time averaged low-frequency ponderomotive force produced by the self-interaction of the high-frequency part of the electric field. One effect of this force is to push the electrons, as we mentioned earlier, out of the regions in the plasma where the electric field has a local maximum.

Using (77), equation (76) becomes

$$\frac{\partial^2 n_\ell}{\partial t^2} - C_s^2 \frac{\partial^2 n_\ell}{\partial x^2} = \frac{1}{16\pi m_i} \frac{\partial^2}{\partial x^2} (|\mathscr{E}|^2). \tag{78}$$

Equation (78) describes the formation of density cavities due to the ponderomotive force associated with the electric field.

Let us now nondimensionalize the various quantities using the ion-plasma period ω_{pi}^{-1}, the Debye length λ_D, n_ℓ using n_0, and E using $4(\pi n_0 K T_e)^{1/2}$. Then equations (71) and (78) become

$$i\varepsilon \mathscr{E}_t + \frac{3}{2} \mathscr{E}_{xx} = \frac{n_\ell}{2} \mathscr{E} \tag{79}$$

$$(n_\ell)_{tt} - (n_\ell)_{xx} = \tfrac{1}{4}(|\mathscr{E}|^2)_{xx} \tag{80}$$

which are called Zakharov's equations. Here, $\varepsilon \equiv (m_e/m_i)^{1/2}$. Equations (79) and (80) have been extensively studied by Gibbons *et al.* [58] who also developed a Lagrangian formalism for this system.

Let us look for a stationary solution of the form

$$n_\ell, \mathscr{E} \sim f(x - Mt) \tag{81}$$

so that equation (80) gives

$$n_\ell = -\frac{|\mathscr{E}|^2}{4(1 - M^2)}. \tag{82}$$

Using (82), equation (79) gives the nonlinear Schrödinger equation

$$i\varepsilon \mathscr{E}_t + \frac{3}{2} \mathscr{E}_{xx} + \frac{1}{8(1 - M^2)} |\mathscr{E}|^2 \mathscr{E} = 0. \tag{83}$$

Put,

$$\mathscr{E}(x, t) = \tilde{\mathscr{E}}(x - Mt) \exp\left[i \left(\frac{\varepsilon M x}{3} - \sigma t \right) \right] \tag{84}$$

so that equation (83) gives

$$\tilde{\mathscr{E}}'' - \frac{2}{3}\left(\frac{M^2}{6} - \varepsilon\sigma \right) \tilde{\mathscr{E}} + \frac{\tilde{\mathscr{E}}^3}{12(1 - M^2)} = 0 \tag{85}$$

from which,

$$\tilde{\mathscr{E}} = 4\left[\left(\frac{M^2}{6} - \varepsilon\sigma\right)(1 - M^2)\right]^{1/2} \times$$

$$\times \operatorname{sech}\left[\sqrt{\frac{2}{3}\left(\frac{M^2}{6} - \varepsilon\sigma\right)}(x - Mt)\right] \tag{86}$$

which represents an envelope soliton. Observe that:

(a) these envelope solitons exist only at subsonic speeds;
(b) these envelope solitons move faster but become smaller and narrower as M increases.

Envelope solitons given by (86) have also been discussed by Karpman [51, 52] and Rudakov [53]. An inverse-scattering solution of Zakharov's equations has not been given, but the latter have been extensively investigated (Gibbons *et al.* [58], Degtyarev *et al.* [166], Rowland *et al.* [167]).

(a) Instability and Collapse of Langmuir Solitons

The envelope solitons of Zakharov's equations are found to be stable to one-dimensional perturbations but unstable to two-dimensional perturbations (Schmidt [50]). In order to demonstrate this, let us consider Zakharov's equations in three dimensions,

$$\nabla \cdot \left(i\frac{\partial \mathscr{E}}{\partial t} + \nabla\nabla \cdot \mathscr{E} - n_\ell \mathscr{E}\right) = 0 \tag{87}$$

$$\frac{\partial^2 n_\ell}{\partial t^2} - \nabla^2 n_\ell = \nabla^2 |\mathscr{E}|^2. \tag{88}$$

Consider a small perturbation about the equilibrium solution given by

$$\begin{aligned} &\mathscr{E}_0 = \mathscr{E}_0(\xi)\hat{\mathbf{i}}_x, \quad \mathscr{E}_0 = [2(1 - M^2)]^{1/2} A \operatorname{sech} A\xi, \\ &n_{\ell_0} = -2A^2 \operatorname{sech}^2 A\xi, \quad \xi = x - Mt \end{aligned} \tag{89}$$

so that,

$$\mathscr{E} = \mathscr{E}_0 + \delta\mathscr{E}\,\hat{\mathbf{i}}_x, \quad n_\ell = n_{\ell_0} + \delta n. \tag{90}$$

One then obtains from equations (87) and (88):

$$\frac{\partial}{\partial x}\left(i\frac{\partial \delta\mathscr{E}}{\partial t} + \nabla^2 \delta\mathscr{E} - n_{\ell_0}\delta\mathscr{E} - \mathscr{E}_0 \delta n\right) = 0 \tag{91}$$

$$\frac{\partial^2 \delta n}{\partial t^2} - \nabla^2 \delta n = \nabla^2(\mathscr{E}_0 \delta\mathscr{E}^* + \mathscr{E}_0^* \delta\mathscr{E}). \tag{92}$$

Putting,

$$\begin{aligned} \delta\mathscr{E} &= F(\xi)e^{ix+\gamma t}\cos k_\perp y \\ \delta n &= N(\xi)e^{ix+\gamma t}\cos k_\perp y. \end{aligned} \tag{93}$$

One obtains from equations (91) and (92):

$$i\gamma F - (1+k_\perp^2)F + \frac{\mathrm{d}^2F}{\mathrm{d}\xi^2} - n_{\ell_0}F - \mathscr{E}_0 N = 0 \tag{94}$$

$$\left(\gamma - M\frac{\mathrm{d}}{\mathrm{d}\xi}\right)^2 N + k_\perp^2 N - \frac{\mathrm{d}^2N}{\mathrm{d}\xi^2} = 2\left(\frac{\mathrm{d}^2}{\mathrm{d}\xi^2} - k_\perp^2\right)[\mathscr{E}_0\,\mathrm{Re}(F)]. \tag{95}$$

Further putting,

$$\begin{aligned} F &= f + ig, \quad z = A\xi, \quad n = N/A \\ \Gamma &= \gamma/A^2, \quad K = k_\perp/A \end{aligned} \tag{96}$$

one obtains from equations (94) and (95):

$$f_{zz} - f + 2f\,\mathrm{sech}^2 z - [2(1-M^2)]^{1/2} n\,\mathrm{sech}\, z - K^2 f = \Gamma g \tag{97}$$

$$g_{zz} - g + 2g\,\mathrm{sech}^2 z - K^2 g = -\Gamma f \tag{98}$$

$$\left(A\Gamma - M\frac{\mathrm{d}}{\mathrm{d}z}\right)^2 n + \left(K^2 - \frac{\mathrm{d}^2}{\mathrm{d}z^2}\right)n$$

$$= -2[2(1-M^2)]^{1/2}\left(K^2 - \frac{\mathrm{d}^2}{\mathrm{d}z^2}\right) f\,\mathrm{sech}\, z. \tag{99}$$

Corresponding to marginal stability, i.e., $\Gamma = 0$, equation (98) has the eigen-solution

$$g = \mathrm{sech}\, z, \quad K = 0. \tag{100}$$

Equations (97) and (99) then give

$$f_{zz} + 6f\,\mathrm{sech}^2 z - f = 0 \tag{101}$$

from which

$$f = -\tanh z\,\mathrm{sech}\, z. \tag{102}$$

Thus, the envelope soliton given by (89) is marginally stable with respect to one-dimensional perturbations.

In order to consider the two-dimensional perturbations, let us write equations (97) and (98) as follows,

$$Lf - K^2 f - [2(1-M^2)]^{1/2} n\,\mathrm{sech}\, z = \Gamma g \tag{103}$$

$$Lg - K^2 g = -\Gamma f \tag{104}$$

where

$$L \equiv \frac{d^2}{dz^2} + 2\,\text{sech}^2 z - 1.$$

Multiplying equations (103) and (104) by φ where

$$L\varphi = 0 \tag{105}$$

or

$$\varphi = \text{sech}\, z$$

and integrating over z, one obtains

$$\Gamma^2 = -K^4 - K^2[2(1-M^2)]^{1/2}\,\frac{\int n\varphi\,\text{sech}\, z\, dz}{\int f\varphi\, dz}. \tag{106}$$

In order to exhibit instability, Schmidt [50] evaluates (106) by considering the limit $A\Gamma \gg 1$ and $\Gamma \gg 1$ which does not seem to be very physical. (Schmidt's work has also been criticized by Infeld and Rowlands [168].) One may accomplish this by simply choosing as trial functions in (91):

$$n = -B\,\text{sech}^2 z, \quad f = \text{sech}\, z, \quad B > 0,$$

B being a constant. Equation (106) then gives

$$\Gamma^2 = -K^4 + K^2[2(1-M^2)]^{1/2}(2B/3). \tag{107}$$

Thus, if B is large enough, an instability results.

It may be mentioned that another approach to investigate the stability of Langmuir solitons is through the Liapunov functional method (Laedke and Spatschek [169, 170]). The stability of Langmuir solitons has also been considered by Katyshev and Makhankov [171], Infeld and Rowlands [172], and Anderson *et al.* [173].

The above instability of the Langmuir soliton is believed to lead to a collapse, i.e., a shrinkage in the physical dimensions, of the solitons accompanied by a corresponding increase in energy density. This result is apparent by exhibiting self-similar solutions to the three-dimensional Zakharov's equations of the form (Zakharov [45])

$$\mathscr{E} \sim \frac{1}{\zeta} f\left(\frac{r}{\zeta}\right), \quad \zeta = \zeta(t) = \sqrt{t_c - t}. \tag{108}$$

The 'self-similar' nature of this solution is apparent because the initial spatial shape $f(r/\zeta(0))$ is preserved at later times, although contracted, since $\zeta(0)$ is replaced by the smaller scale sizes $\zeta(t)$, which eventually go to zero. The solution $\mathscr{E}$ is therefore similar to itself at later times, except that it becomes narrower, and also larger, because of the factor $1/\zeta$ in front of f. The spatial collapse may also be deduced by a 'virial theorem' for the time evolution of the wave-packet averaged width of an initial wave packet in the adiabatic approximation (Goldman and Nicholson [46]). The spatial collapse was also revealed by the numerical solutions of Zakharov's equations in two dimensions for an initially localized pulse by Pereira *et al.* [47] who found that the self-similarity of the collapse was anisotropic, and in three dimensions by Goldman *et al.* [174], and Hafizi and Goldman [175]. During this collapse, when the scale length approaches the Debye length, strong Landau damping follows and the wave energy is dissipated into particle energy.

(b) Nearly Sonic Langmuir Solitons

Observe that the envelope soliton solution given by (71) of Zakharov's equations is not valid when $M \approx 1$. In order to develop a solution valid for the latter case, one has to include the nonlinearities in the ion motion (Makhankov [59], and Nishikawa *et al.* [60]). Makhankov [59] and Nishikawa *et al.* [60] accounted for the ion nonlinearities by assuming charge neutrality in the low-frequency plasma response, and the low-frequency electron-density perturbation they obeyed, respectively, a Boussinesq equation and a Korteweg—de Vries equation with the ponderomotive force as the driving term, while the Langmuir field continued to obey the nonlinear Schrödinger equation. Further study of these approaches have been made by Shivamoggi [176]. Rao and Varma [61] considered a fully-nonlinear system of equations without an assumption of charge neutrality in the low-frequency plasma response and gave a solution which shows a smooth transition from a single-hump soliton solution of Zakharov [45] to a double-hump soliton solution of Nishikawa *et al.* [60] as the Mach number M of the soliton increases. We shall treat this problem below by using a simpler method like the one given in Shivamoggi [62].

One has for the ions (see Section 3)

$$\frac{n_{i_\ell}}{n_0} = \frac{1}{\sqrt{1 - 2\phi/M^2}} - 1 \tag{108}$$

where ϕ is the ambipolar-field potential.

Assuming that the electrons respond adiabatically to the ponderomotive force and are distributed according to the Boltzmann law in the low-frequency motions, one has

$$\frac{n_{e_\ell}}{n_0} = \exp\left(\phi - \frac{|E|^2}{4}\right) - 1. \tag{109}$$

The fully nonlinear system of equations for this problem now are (79) and (109), i.e.,

$$i\varepsilon E_t + \frac{3}{2} E_{xx} = \frac{1}{2}\left[\exp\left(\phi - \frac{|E|^2}{4}\right) - 1\right] E \tag{110}$$

and Poisson's equation for the space charge field ϕ, using (108) and (109), i.e.,

$$\phi_{xx} = -\frac{1}{\sqrt{1 - 2\phi/M^2}} + \exp\left(\phi - \frac{|E|^2}{4}\right). \tag{111}$$

In the linearized limit equations (110) and (111) display the usual parametric excitation of ion-acoustic waves by Langmuir waves. The nonlinear solutions correspond to the special saturated states of such parametric excitations.

Let us again look for stationary solutions of the form

$$E(x,t) = \mathscr{E}(\xi)\exp i\left(\frac{\varepsilon Mx}{3} + \sigma t\right)$$
$$\phi(x,t) = \phi(\xi) \tag{112}$$

where

$$\xi \equiv x - Mt.$$

Equations (110) and (111) then become

$$\mathscr{E}'' - \Delta\mathscr{E} = \frac{1}{3}\left[\exp\left(\phi - \frac{\mathscr{E}^2}{4}\right) - 1\right]\mathscr{E} \tag{113}$$

$$\phi'' = -\frac{1}{\sqrt{1 - 2\phi/M^2}} + \exp\left(\phi - \frac{\mathscr{E}^2}{4}\right). \tag{114}$$

Equations (113) and (114) correspond to the 'Hamiltonian'

$$H = \tfrac{3}{4}\mathscr{E}'^2 + (1 - 3\Delta)\mathscr{E}^2 - \tfrac{1}{2}\phi'^2 + \exp(\phi - \mathscr{E}^2/4)$$
$$+ M^2(1 - 2\phi/M^2)^{1/2}.$$

Observe that H has an indefinite character. Therefore, the motion of the effective particle represented by H can become unbounded. For the present system, this implies that the resonant coupling of the two waves takes the ion waves over its potential hill into the ion-sheath solution (see Figure 5.3), which represents unbounded motions. Here,

$$v(\phi) = (1 - 3\Delta)\frac{\mathscr{E}^2}{4} + \exp\left(\phi - \frac{\mathscr{E}^2}{4}\right) + M^2\left(1 - \frac{2\phi}{M^2}\right)^{1/2}.$$

Figure 5.2 also shows nonlinear periodic solutions, and solitary-wave solutions which correspond to an infinite period. The ion-sheath solutions first discussed by Bohm [177] correspond to large negative ϕ with

$$|\phi''| \sim \frac{M}{\sqrt{2}}(|\phi|)^{-1/2}$$

from equation (114), for $\mathscr{E} = 0$. Thus, $|\phi| \sim \xi^{4/3}$. This solution corresponds to cases wherein the electrons cannot catch up with the ions and one thus obtains an ion-rich sheath, and the electric field is unable to penetrate the plasma.

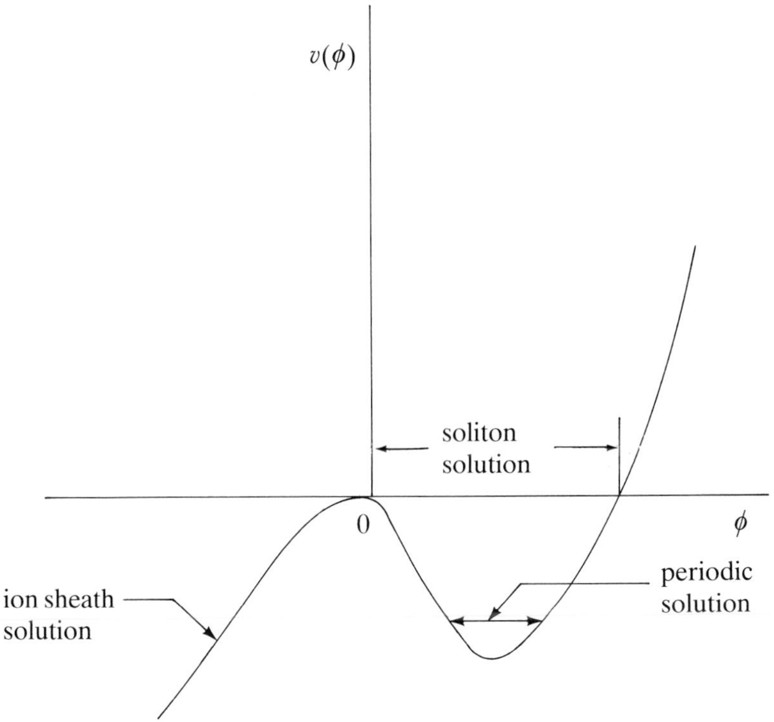

Figure 5.3. Quasi-potential diagram for coupled Langmuir — ion acoustic waves.

Upon expanding the exponentials and the radical, and keeping the terms to $0(\xi^4)$, equations (113) and (114) become

$$\mathscr{E}\mathscr{E}'' - \Delta\mathscr{E}^2 = \frac{1}{3}\left[\phi - \frac{\mathscr{E}^2}{4} + \frac{1}{2}\left(\phi^2 - \frac{\phi\mathscr{E}^2}{2} + \frac{\mathscr{E}^4}{16}\right)\right]\mathscr{E}^2 \tag{115}$$

$$\phi'' - \left(\frac{M^2-1}{M^2}\right)\phi + \frac{\mathscr{E}^2}{4} - \frac{1}{2}\left[\left(1 - \frac{3}{M^4}\right)\phi^2 - \frac{\phi\mathscr{E}^2}{2} + \frac{\mathscr{E}^4}{16}\right] = 0. \tag{116}$$

Put,

$$\frac{\mathscr{E}^2}{4} = a\phi + b\phi^2 \tag{117}$$

so that equations (115) and (116) become

$$(a+2b\phi)\phi'' + \frac{1}{2}\phi'^2\left[4b - \frac{(a+2b\phi)^2}{(a\phi+b\phi^2)}\right] - 2\Delta(a\phi+b\phi^2)$$

$$= \tfrac{2}{3}\{\phi - (a\phi+b\phi^2) + \tfrac{1}{2}[\phi^2 - 2\phi(a\phi+b\phi^2) + (a\phi+b\phi^2)]\} \times (a\phi+b\phi^2) \tag{118}$$

$$\phi'' - \left(\frac{M^2-1}{M^2}\right)\phi + (a\phi+b\phi^2) - \frac{1}{2}\left[\left(1-\frac{3}{M^4}\right)\phi^2 - 2\phi(a\phi+b\phi^2) + (a\phi+b\phi^2)^2\right] = 0. \tag{119}$$

Integrate equation (119) so that

$$\frac{\phi'^2}{2} - \left(\frac{M^2-1}{M^2}\right)\frac{\phi^2}{2} + \left(\frac{a\phi^2}{2} + \frac{b\phi^3}{3}\right) - \frac{1}{2}\left[\left(1-\frac{3}{M^4}\right)\frac{\phi^3}{3} - 2\left(\frac{a\phi^3}{3} + \frac{b\phi^4}{4}\right) + \left(\frac{a^2\phi^3}{3} + \frac{ab\phi^4}{2} + \frac{b^2\phi^5}{5}\right)\right] = 0. \tag{120}$$

Using (119) and (120), equation (118) becomes

$$(a+b\phi)(a+2b\phi)\left[\left(\frac{M^2-1}{M^2}\right) - (a+b\phi) + \frac{1}{2}\left\{\left(1-\frac{3}{M^4}\right)\phi - 2(a\phi+b\phi^2) + \phi(a+b\phi)^2\right\}\right]$$

$$+\left[\frac{1}{2}\left(\frac{M^2-1}{M^2}\right) - \left(\frac{a}{2} + \frac{b\phi}{3}\right) + \frac{1}{2}\left\{\frac{\phi}{3}\left(1-\frac{3}{M^4}\right) - 2\left(\frac{a\phi}{3} + \frac{b\phi^2}{4}\right) + \left(\frac{a^2\phi}{3} + \frac{ab\phi^2}{2} + \frac{b^2\phi^3}{5}\right)\right\}\right] \times$$

$$\times [4b(a\phi+b\phi^2) - (a+2b\phi)^2] - 2\Delta(a+b\phi)^2$$

$$= \tfrac{2}{3}\{\phi - (a\phi+b\phi^2) + \tfrac{1}{2}[\phi^2 - 2\phi(a\phi+b\phi^2) + (a\phi+b\phi^2)]\}(a+b\phi)^2 \tag{121}$$

from which, equating the coefficients of equal powers of ϕ, one obtains

$$\frac{1}{2}\left[\left(\frac{M^2-1}{M^2}\right)-a\right]a^2-2\Delta a^2=0 \tag{122}$$

$$a^2\left[-b+\frac{1}{2}\left\{\left(1-\frac{3}{M^4}\right)-2a+a^2\right\}\right]$$

$$+3ab\left[\left(\frac{M^2-1}{M^2}\right)-a\right]$$

$$-a^2\left[-\frac{b}{3}+\frac{1}{2}\left\{\frac{1}{3}\left(1-\frac{3}{M^4}\right)-\frac{2a}{3}+\frac{a^2}{3}\right\}\right]$$

$$-4\Delta ab=\frac{2}{3}(1-a)a^2$$

where,

$$\Delta\equiv\frac{2}{3}\left(\frac{\varepsilon^2M^2}{6}+\varepsilon\sigma\right). \tag{123}$$

From (123), one obtains:

$$a=\left(\frac{M^2-1}{M^2}\right)-4\Delta \tag{124}$$

$$b=\frac{-\frac{1}{3}\left(1+\frac{3}{M^4}\right)\left[\left(\frac{M^2-1}{M^2}\right)-4\Delta\right]+\frac{1}{3}\left[\left(\frac{M^2-1}{M^2}\right)-4\Delta\right]^3}{-8\Delta+\frac{2}{3}\left[\left(\frac{M^2-1}{M^2}\right)-4\Delta\right]}. \tag{125}$$

Using (124), equation (119) becomes

$$\phi''-4\Delta\phi+\left[b-\frac{1}{2}\left\{\left(1-\frac{3}{M^4}\right)-2a+a^2\right\}\right]\phi^2$$

$$+b(1-a)\phi^3=0. \tag{126}$$

On integrating once and using the boundary conditions:

$$\xi\Rightarrow\infty:\quad \phi,\phi'\Rightarrow 0 \tag{127}$$

equation (86) gives

$$\frac{\phi'^2}{2} = \frac{b(a-1)}{4}\,\phi^2\left[\phi^2 + \frac{4}{3}\,\frac{\{b - \frac{1}{2}[(1-3/M^4) - 2a + a^2]\}}{b(1-a)}\,\phi - \frac{8\Delta}{b(1-a)}\right]$$

$$= \frac{b(a-1)}{4}\,\phi^2(\phi-\beta_1)(\phi-\beta_2) \tag{128}$$

where

$$\beta_{1,2} = \frac{-m \mp \sqrt{m^2 - 4n}}{2}$$

$$n = -\frac{8\Delta}{b(1-a)} \tag{129}$$

$$m = \frac{4}{3}\,\frac{\{b - \frac{1}{2}[(1-3/M^4) - 2a + a^2]\}}{b(1-a)}.$$

Thus,

$$\frac{d\phi}{\phi\sqrt{(\phi-\beta_1)(\phi-\beta_2)}} = \sqrt{\frac{b(a-1)}{2}}\,d\xi. \tag{130}$$

Putting,

$$z^2 = \frac{\beta_1}{\beta_2}\,\frac{\phi-\beta_2}{\phi-\beta_1} \tag{131}$$

equation (130) becomes

$$\frac{1}{2}\sqrt{\frac{b(a-1)}{2}}\int d\xi = \frac{1}{\sqrt{\beta_1\beta_2}}\int\frac{dz}{1-z^2} \tag{132}$$

from which,

$$z = \tanh\sqrt{\frac{b(a-1)}{2}\,\beta_1\beta_2}\;\xi. \tag{133}$$

Noting from (88) that

$$\frac{b(a-1)}{2}\,\beta_1\beta_2 = 4\Delta \tag{134}$$

and using (133), (131) gives

$$\phi = \frac{\beta_1\beta_2\,\mathrm{sech}^2\sqrt{\Delta}\;\xi}{\beta_1 - \beta_2\tanh^2\sqrt{\Delta}\;\xi} \tag{135}$$

which may be approximated as

$$\phi \approx \frac{6\Delta}{b + 1/M^4} \operatorname{sech}^2 \sqrt{\Delta}\, \xi. \tag{136}$$

Using (124), (125), and (136), (117) has been evaluated and plotted in Figure 5.4. Observe that this solution shows a smooth transition from a single hump solution of Zakharov [45] to a double-hump soliton solution of Nishikawa *et al.* [60] as M increases.

In order to recover Zakharov's [45] solution from the above solution, note that the former refers to cases with

(i) $M \neq 1$;
(ii) charge-neutrality prevalent in the low-frequency motion;
(iii) ion-response linear.

Equation (117) then gives

$$\frac{\mathscr{E}^2}{4} = \left(\frac{M^2 - 1}{M^2}\right)\phi \tag{137a}$$

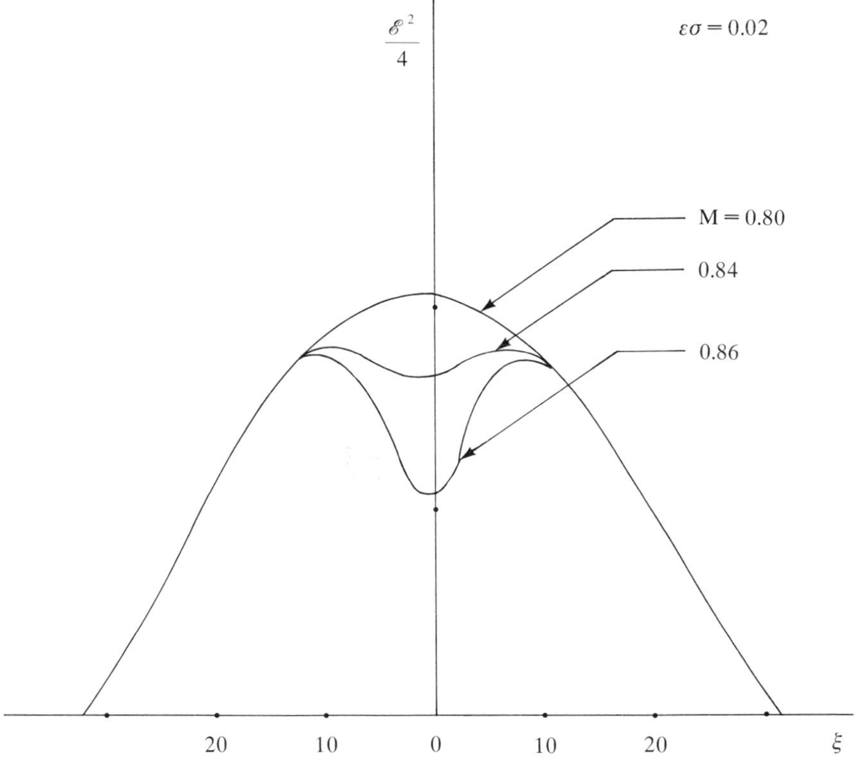

Figure 5.4. Langmuir soliton.

and (109) gives

$$n_{e_\ell} = \phi - \frac{\mathscr{E}^2}{4} = -\frac{\mathscr{E}^2}{4(1-M^2)} \tag{137b}$$

which leads to the single-hump soliton (86).

It should be noted that for very large amplitudes departures from the Boltzmann distribution and reflection of electrons by the negative ambipolar electrostatic potential occur which would eliminate the existence of symmetric solutions.

To recapitulate, in the interaction of waves with a plasma, the waves redistribute the plasma particles via the ponderomotive effects. This in turn influences the evolution of the waves. In the foregoing, the analysis of ponderomotive effects has been *ad hoc*. In the derivation of the ponderomotive potential, on the one hand, the cold-plasma limit is used, but on the other hand, the response to the ponderomotive potential is taken to be given by the Boltzmann factor — a result valid for a hot plasma! This is partly reasonable as long as the phase velocity of the high-frequency waves is large, and the characteristic velocity of the slowly-varying perturbations is small. However, a very basic question remains. Why should the plasma respond to low-frequency ponderomotive potential via the Boltzmann factor? After all, the ponderomotive potential is not identical to a real potential. Besides, the plasma in the presence of a large-amplitude wave is far from equilibrium. These questions have been discussed by Cary and Kaufmann [178].

Experimental observations of self-focusing and modulational instability of a laser beam in a nonlinear medium were made by Garmire *et al.* [64] and Campillo *et al.* [65]. Garmire *et al.* [64] gave direct observations of the evolution of beam trapping in CS_2 in the simplest cylindrical mode. Figure 5.5 shows the magnified images of the beam profile as it reflects off the glass plates placed along the beam length. It was found that after a certain distance a steady-state condition was approached consisting of a bright filament, 100 μ in diameter. Simple diffraction would double the size of such a beam in the distance to the next glass plate, proving that there was indeed beam trapping, i.e., propagation of a beam without spreading by ordinary diffraction. An essential feature of self-trapping in optical beams is the increase in refractive index in the trapped region. The resulting slower velocity of the trapped beam causes an accumulation of the phase of a trapped ray relative to those which are untrapped. This phase change across the beam profile was detected. Campillo *et al.* [65] made experimental observations of the transverse spatial periodic break-up of an optical beam due to self-focusing. Each focal spot evolved from a zone with well defined dimensions.

Observations of density cavities and associated localized Langmuir wave envelopes were made by Ikezi *et al.* [66, 67], Kim *et al.* [68], Cheung *et al.* [69], Wong and Quon [70], and Leung *et al.* [71]. Ikezi *et al.* [66] made observation of a self-modulation of a strong high-frequency electric field (pump field) and the associated density depression or plasma cavity which trapped the electron plasma wave and moved approximately with the ion-acoustic speed. Since this

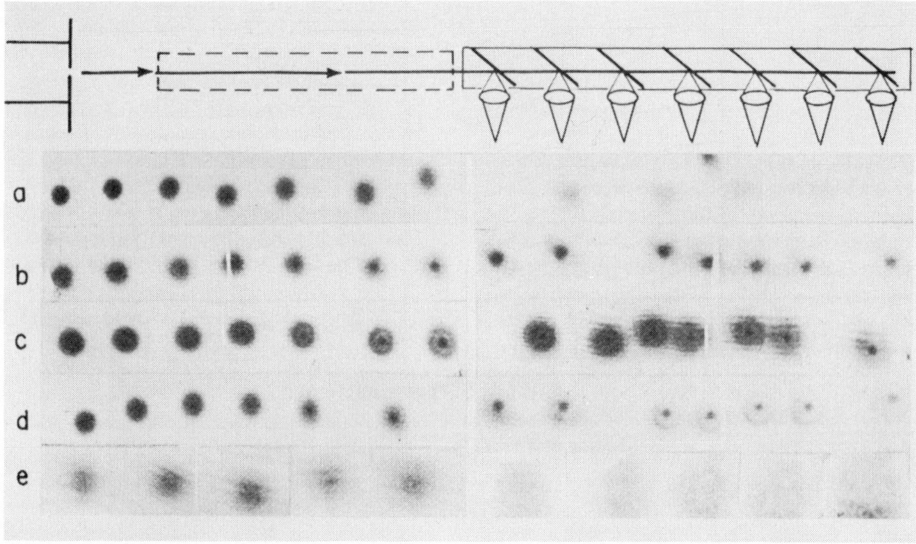

Figure 5.5. Evolution of beam trapping in CS_2. Left: without dashed cell; right: dashed cell adds 25 cm path length. (a) Gas laser control; (b), (c), and (d), beam trapping at increasing power; (e) 1-mm pinhole. (Due to Garmire *et al.* [64], by courtesy of The American Physical Society).

phenomenon was observed in the overdense region of the nonuniform plasma where the electron plasma frequency ω_{p_e} is greater than the pump frequency ω_0, it was clearly separated from the usual decay instability which was also observed at a lower threshold, but in the underdense region ($\omega_0 > \omega_{p_e}$). The growth of the modulational instability was observed by applying the stepwise pump power starting at time $t = 0$. The typical distribution of the plasma density and of the high-frequency electric field along the direction of the pump electric field are shown in Figure 5.6 for every 1 μsec time interval after the pump has been turned on. The plasma density started decreasing at a place near the critical layer where $\omega_0 = \omega_{p_e}$; then the density depression expanded toward both the overdense ($\omega_{p_e} > \omega_0$) and the underdense ($\omega_{p_e} < \omega_0$) regions with a velocity approximately equal to the ion-acoustic velocity. The expanding front eventually formed a narrow dip in the overdense region and when this dip moved away another dip arose at the place where the first dip appeared. This process was repeated and an oscillating structure was eventually formed. Figure 5.6b shows the correlation between the intensity of the high-frequency electric field and the density perturbation. At the position where the density had dropped, the electric field intensity E^2 increased to about 5 times the unperturbed value. The envelope of the high-frequency electric field propagated together with the density cavity.

In another experiment Ikezi *et al.* [67] investigated the parametric instabilities associated with large-amplitude waves excited by the electron-beam-plasma instability. It was found that the trapping of the electron-plasma wave by the

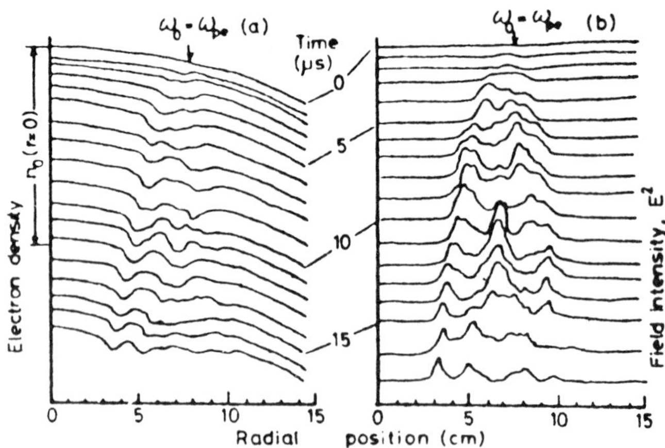

Figure 5.6. Distributions of the density perturbation (a) and the high-frequency field intensity (b). A stepwise pump starts at $t = 0$. Pump power = 20 W. (Due to Ikezi *et al.* [66], by courtesy of Journal of The Physical Society of Japan).

ion-acoustic wave modifies the convective nature of the beam-plasma instability to a temporal one.

Wong and Quon [70] gave experimental evidence on the spatial collapse of large-amplitude plasma waves driven by an electron beam in a uniform plasma. Electron-plasma waves became unstable and were observed to be backscattered by stationary ion-density perturbations to form standing waves. The ponderomotive force due to the standing wave started to dig the density cavity further and to trap the electric field which eventually developed into sharply localized (of the order of 10 Debye lengths) field structures. Figure 5.7 shows a typical measurement of the stationary ion-density perturbation n_1/n_0 and the standing-wave-like amplitude modulation of the total high-frequency electric field $\langle E_T^2 \rangle$ in the steady state, where $E_T = E_0 + E_2$, E_0 is the beam-driven pump field and E_2 is the field caused by the coupling between the beam-amplified wave and the stationary spatial density variation, and $\langle \rangle$ represents averaging over one high-frequency wave period. For pump field above the threshold value both $\langle E_T^2 \rangle$ and n_1/n_0 grow in time with the same growth rate. When the pump field was increased further, the initial growth was followed by increased sharpening of the modulation in $\langle E_T^2 \rangle$, demonstrating the collapse of plasma waves (Figure 5.8) into localized regions of 10 Debye lengths. Significant fractions of beam electrons were found to be scattered in velocity space by these stationary spikes of high-frequency field, signifying the dissipation of wave energy into particle energy due to Landau damping that becomes effective at such small length scales. Cheung *et al.* [69] also gave experimental evidence on the spatial collapse of a Langmuir wave excited by a cold-electron beam. The Langmuir wave turning unstable led to the creation of a density cavity. New unstable electrostatic waves were subsequently excited by the beam inside the cavity at the lower plasma frequency corresponding to the fallen density of the cavity. These

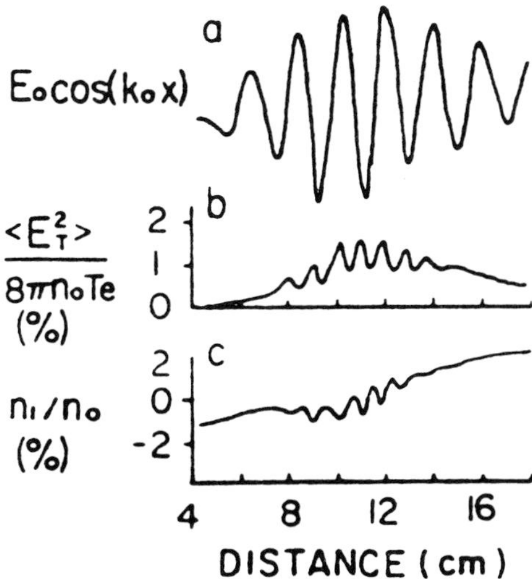

Figure 5.7. Excitation of the stationary density perturbation. (a) Interferometer trace of the electron-plasma wave driven by the beam $f_0 = 200$ MHz, $k_0 \sim 3$ cm^{-1}, $E_0 \sim 2.5$ V/cm. (b) Time averaged $\langle E_T^2 \rangle$ measured by a crystal diode as a function of distance, showing the standing-wave component of E_T; E_T is the total field in plasma. (c) The stationary ion-density perturbation, $k_1 \sim 6$ cm^{-1}, $n_1/n_0 \sim 1\%$. (Due to Wong and Quon [70], by courtesy of The American Physical Society.)

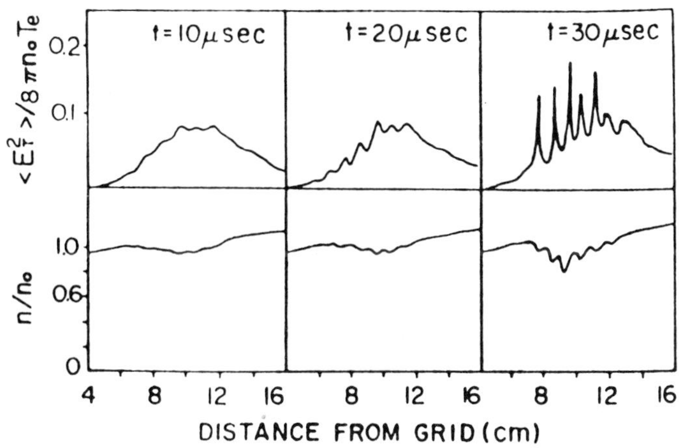

Figure 5.8. Temporal development of the high-frequency wave $\langle E_T^2 \rangle$ (top traces) and ion-density perturbation n_1/n_0 (bottom traces) for $E_0 \sim 5$ V/cm, at successive times after turn on of the pump wave. (Due to Wong and Quon [70], by courtesy of The American Physical Society).

waves, which were evanescent outside the cavity were trapped and driven to spatially localized structures of wave intensity E^2.

An efficient way of generating localized electric field structures was used in the experiment of Leung *et al.* [71]. In the experiment of Wong and Quon [70] the standing wave pattern preceding the collapse stage relied on the parametrically-excited backward electron plasma wave. Leung *et al.* [71], by contrast, generated the forward and backward electron plasma waves simultaneously using two oppositely-propagating electron beams interacting with the plasma. As one did not need to go through the parametric decay to get the backward-propagating wave and have it grow to the same level as the forward-propagating wave, Leung *et al.* [71] found that the time required for the electron plasma waves to evolve into the collapse stage was much shorter than that in the case of one beam.

A characteristic feature of the present problem is that the initial wave (or pump wave) modifies the properties of the plasma medium in such a way that the frequencies of the excited waves or perturbations are shifted from the values these waves would have in the absence of the pump wave. The nonlinear theory of Langmuir solitons and Langmuir-wave collapse discussed above considers only the resultant envelope of the pump wave and excited high-frequency waves. Bingham and Lashmore-Davies [179, 180] formulated the problem, on the other hand, in such a way that instead of averaging over the pump and excited waves, the pump and excited waves were distinguished from one another throughout the interaction — this procedure showed the four-wave nature of the interaction.

(c) Effect of Finite Spectral-Width on the Modulational Instability of Langmuir Waves

The above treatments were mostly restricted to the case of monochromatic Langmuir waves. However, the bandwidth of an initially ‘narrow’ spectrum may eventually broaden, either as a result of the modulational instability process itself, or as the result of other weaker nonlinear wave-wave interaction processes. Thornhill and ter Haar [77] conjectured that a finite bandwidth will not seriously affect the instability if the frequency spread of the wave-packet is less than the growth rate of the modulational instability and the wavenumber width Δk is less than the wavenumber at which the maximum growth occurs. The effect of finite spectral-width on the modulational instability of Langmuir waves was investigated by Breizman and Malkin [181] and Bhakta and Majumdar [182] who assumed the spectral width to be small compared with the wavenumber of the carrier wave.

Let us introduce the two-point correlation function

$$P(x_1, x_2; t) = \langle E(x_1, t) E^*(x_2, t) \rangle \tag{138}$$

where $\langle \ldots \rangle$ denotes the ensemble average. The equation for P can now be obtained from equation (79):

$$i \frac{\partial P}{\partial t} + \left(\frac{\partial^2}{\partial x_1^2} - \frac{\partial^2}{\partial x_2^2} \right) P = [n(x_1, t) - n(x_2, t)] P. \tag{139}$$

Upon introducing,

$$x = \tfrac{1}{2}(x_1 + x_2), \quad \xi = x_1 - x_2 \tag{140}$$

equation (139) becomes:

$$i\,\frac{\partial P}{\partial t} + 2\,\frac{\partial^2 P}{\partial \xi\,\partial x} = [n(x + \tfrac{1}{2}\xi, t) - n(x - \tfrac{1}{2}\xi, t)]P. \tag{141}$$

Fourier transforming with respect to ξ, and introducing the power spectral function:

$$F(x, p, t) = \frac{1}{2\pi}\int_{-\infty}^{\infty} P(x + \tfrac{1}{2}\xi, x - \tfrac{1}{2}\xi; t)\,e^{-ip\xi}\,\mathrm{d}\xi \tag{142}$$

equation (141) becomes

$$\frac{\partial F}{\partial t} + 2p\,\frac{\partial F}{\partial x} - \sum_{r=1}^{\infty} \frac{2(-1)^{r-1}}{2^{2r-1}(2r-1)!}\left(\frac{\partial^{2r-1}F}{\partial p^{2r-1}}\right)\left(\frac{\partial^{2r-1}n}{\partial x^{2r-1}}\right) \tag{143}$$

where we have Taylor-expanded the term in the bracket on the right-hand side in equation (141) about x.

The energy density of the Langmuir wave field is given by:

$$|E(x, t)|^2 = \int_{-\infty}^{\infty} F(x, p, t)\,\mathrm{d}p$$

so that the relation between the density fluctuation $n(x, t)$ and the spectral distribution is provided by equation (80):

$$\frac{\partial^2 n}{\partial t^2} - \frac{\partial^2 n}{\partial x^2} - \frac{\partial^2}{\partial x^2}\int_{-\infty}^{\infty} F(x, p, t)\,\mathrm{d}p = 0. \tag{144}$$

Let us now consider a perturbation about the stationary state $F(x, p, t) = F_0(p)$, which corresponds also to $n = 0$:

$$\begin{aligned} F(x, p, t) &= F_0(p) + F_1(p)e^{i(kx-\omega t)} + \cdots \\ n(x, t) &= n_1\,e^{i(kx-\omega t)} + \cdots. \end{aligned} \tag{145}$$

Using (145) and linearizing in F_1 and n_1, equations (143) and (144) give:

$$i(-\omega + 2pk)F_1 - \sum_{r=1}^{\infty} \frac{2(-1)^{r-1}}{2^{2r-1}(2r-1)!}\left(\frac{\partial^{2r-1}F_0}{\partial p^{2r-1}}\right)(ik)^{2r-1}n_1 = 0 \tag{146a}$$

$$(\omega^2 - k^2)n_1 - k^2\int_{-\infty}^{\infty} F_1(p)\,\mathrm{d}p = 0. \tag{146b}$$

Using (146b), and summing the series, (146a) gives:

$$(2pk - \omega)F_1 - [F_0(p + \tfrac{1}{2}k) - F_0(p - \tfrac{1}{2}k)] \times$$
$$\times \frac{k^2}{\omega^2 - k^2} \int_{-\infty}^{\infty} F_1(p)\,dp = 0 \tag{147}$$

from which the dispersion relation immediately follows:

$$1 - \frac{k}{2(\omega^2 - k^2)} \int_{-\infty}^{\infty} \frac{F_0(p + \tfrac{1}{2}k) - F_0(p - \tfrac{1}{2}k)}{p - \omega/2k}\,dp = 0. \tag{148}$$

For a Gaussian spectrum

$$F_0(p) = \frac{F_0}{\sqrt{2\pi}}\, e^{-p^2/2\sigma^2}.$$

Equation (148) was numerically evaluated by Bhakta and Majumdar [182] and the result is shown in Figure 5.9. Observe that the maximum growth rate diminishes as spectral-width σ increases, and for a sufficiently large spectral-width the instability can be suppressed. Physically, an increase in spectral-width leads to decorrelation of the phases of the wave envelope which brings about stabilization of the wave train.

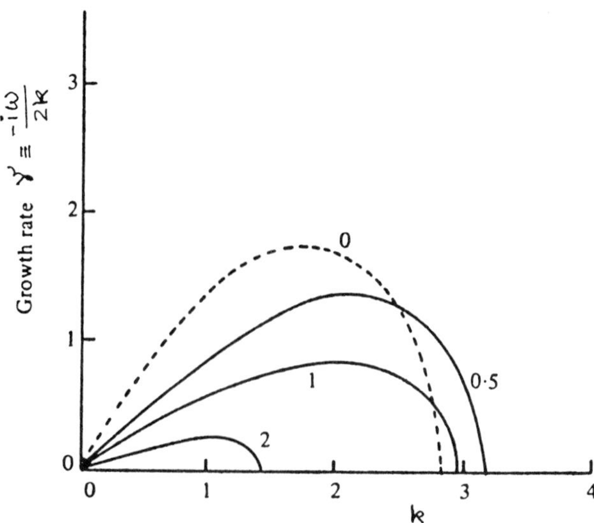

Figure 5.9. Growth rate γ plotted against wavenumber for the Gaussian spectrum given by (13). γ and F_0 are respectively in units of $(\frac{2}{3}\omega_{p_e} m_e/m_i)^{1/2}$ and $8(\pi n_0 T_e m_e/m_i)^{1/2}$; k and σ are in units of $\frac{2}{3}\lambda_D^{-1}(m_e/m_i)^{1/2}$. $F_0 = 4$. Values of σ^2 are shown against the curves. (Due to Bhakta and Majumdar [182], by courtesy of Cambridge University Press.)

(d) Strong Langmuir Turbulence

Kingsep *et al.* [72] formulated a model for strong Langmuir turbulence which consisted not of a Fourier superposition of plane waves, but of a collection of one-dimensional Langmuir solitons, localized in space with random positions and an *ad hoc* distribution of amplitudes. They found a spectrum porportional to k^{-2} and noted that near $k \approx \lambda_D^{-1}$ there would occur a strong absorption of wave energy by Landau damping. Galeev *et al.* [73, 74] gave a more accurate model of strong Langmuir turbulence, valid in three dimensions and based on an 'ideal gas' of self-similarly collapsing solitons. They found an isotropic, stationary spectrum proportional to $k^{-5/2}$ at intermediate wavenumbers and noted that collapse provides the mechanism for energy transfer to high k. Dissipation of wave energy is provided by quasi-linear coupling to electrons at short scale sizes associated with the collapsed end states. This theory was refined further by Pelletier [183].

Experimental observation of such a model of strong Langmuir turbulence characterized by rapidly collapsing wave packets and local density inhomogeneities was indeed made by Cheung and Wong [75] in their pulsed beam experiment wherein a single Langmuir wave packet formed and collapsed three-dimensionally in a self-similar manner. The transverse contraction rate of the wave packet was observed to follow theoretical predictions until other nonlinear effects become important at small scale lengths. This instability was excited in a beam-plasma system, and since the most unstable wave has a frequency slightly less than the ambient plasma frequency, the nonlinear development of this instability led to a rapid axial and transverse localization of high-frequency plasma waves trapped inside collapsing density cavities. Spatial and temporal evolutions of the electric field and density cavity driving the collapse process were obtained. Figure 5.10 shows the time-averaged temporal evolution of the electric field at selected times during the collapse process. At $\omega_{p_i}t \approx 5.7$, the initial field structure had an axial width (full width at half maximum) of $\Delta z/\lambda_D \approx 500$ and a radial width of $\Delta r/\lambda_D \approx 150$. For $\omega_{p_i}t \leqslant 12$, the spatial contraction was mainly along the beam direction. At $\omega_{p_i}t \approx 32.9$, the field structure started to break up, and the spatial widths contracted to $\Delta z/\lambda_D \approx 160$ and $\Delta r/\lambda_D \approx 93$. At $\omega_{p_i}t \approx 55.8$, the field had contracted both axially and transversely to $\Delta z/\lambda_D \approx 42$ and $\Delta r/\lambda_D \approx 42$ and $\Delta r/\lambda_D \approx 39$. Soon after this time, Δr stopped decreasing and remained approximately constant at $\Delta r/\lambda_D \approx 43$. Both the increase of the field intensity and the contraction of the axial width thereafter proceeded at a much slower rate. When $\omega_{p_i}t \approx 78$, the axial width was $\Delta z/\lambda_D \approx 20$ and the field structure had an elongated, pancake shape with the ratio of transverse to axial width approximately equal to 2. Theoretically, the contraction rate was deduced from equations (79), (80) by Galeev *et al.* [73] to be $L \propto (t_0 - t)^{2/d}$ where L is the width of the caviton field, d is the number of spatial dimensions, and t_0 was defined as the time it takes the field to collapse to a singular point. For a three-dimensional Langmuir collapse, $d = 3$, and the contraction rate becomes $L \propto (t_0 - t)^{2/3}$. The measured contraction rates of the spatial widths are plotted in Figure 5.11. The radial contraction rate was

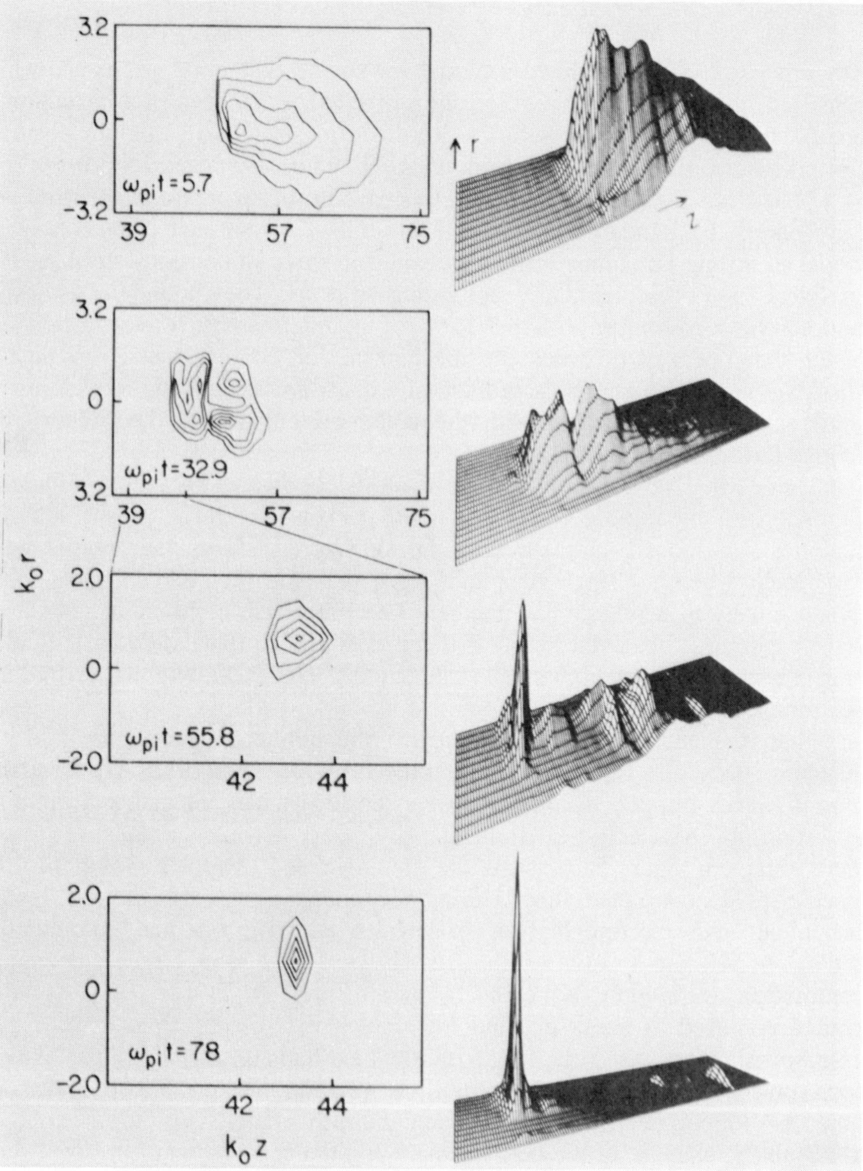

Figure 5.10. Temporal evolution of the collapsing wave packet. Two-dimensional contours and three-dimensional views of the field intensity, $E^2(r, z)$, are shown at $\omega_{p_i}t \simeq 5.7$ 32.9, 55.8, and 78, respectively. The contours are in equal increments, with the outermost contour at 0.35 of the peak intensity at each time. The beam extends radially to $k_0 r \simeq \pm 3$. (Due to Cheung and Wong [75], by courtesy of The American Institute of Physics.)

observed to scale approximately as $\Delta r/\lambda_D$ and $t^{2/3}$. The axial-width data points were more scattered and the contraction was roughly linear in time. By extrapolating the contraction rates before $\omega_{p_i}t < 58$, t_0 was found to be at $\omega_{p_i}t \approx 62.5$. After $\omega_{p_i}t > 58$, the contraction in both directions was signifi-

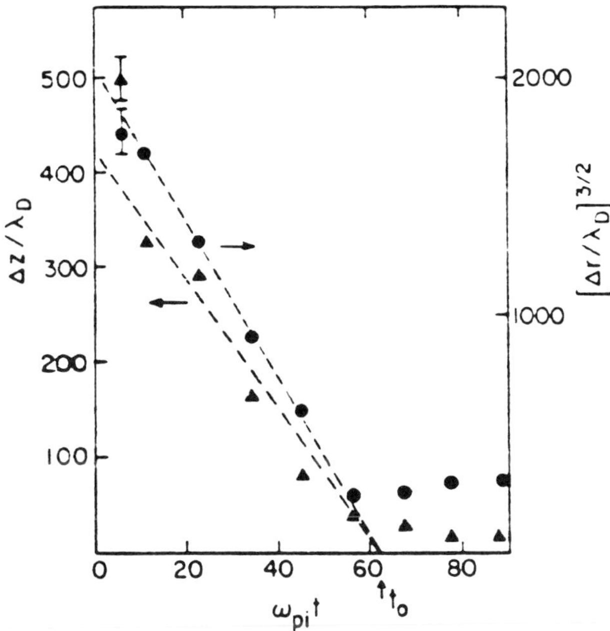

Figure 5.11. Contraction rates of the spatial widths of the collapsing wave packet. Axial widths (closed triangles) are in units of $\Delta z/\lambda_D$ and the radial widths (closed circles) are in units of $(\Delta r/\lambda_D)^{3/2}$. Extrapolation of the contraction rates (dashed lines) determines t_0. (Due Cheung and Wong [75], by courtesy of The American Institute of Physics.)

cantly slower and deviated from theory. At such small-scale lengths, other nonlinear effects become important and must be taken into account.

Another interesting result found by Cheung and Wong [75] was that the number of intense field spikes formed along the beam path driving the collapse process depended on the beam density $\eta = n_b/n_0$. At low beam density ($\eta \leqslant 0.6\%$), a single collapsing wave packet was formed. At high beam density, multiple collapsing wave packets were formed both along and across the beam path. Figure 5.12 shows the dependence of the electric field structures and the corresponding density cavities as a function of the beam density η. If the beam density n_b is kept constant and instead the beam velocity V_b is varied the results observed were qualitatively the same as above (Cheung, personal communication, 1987). This result appears to corroborate the theoretical result discussed in Section (ib) showing a transition of a single-humped soliton to a double-humped one as the speed of propagation increases.

Excellent theoretical accounts of strong Langmuir turbulence have been given by Rudakov and Tsytovich [76], and Thornhill and ter Haar [77].

(ii) Modulational Stability of Ion-Acoustic Waves

Modulational stability of ion-acoustic waves has been studied by Shimizu and Ichikawa [78], Kakutani and Sugimoto [79], Chan and Seshadri [80], Infeld and

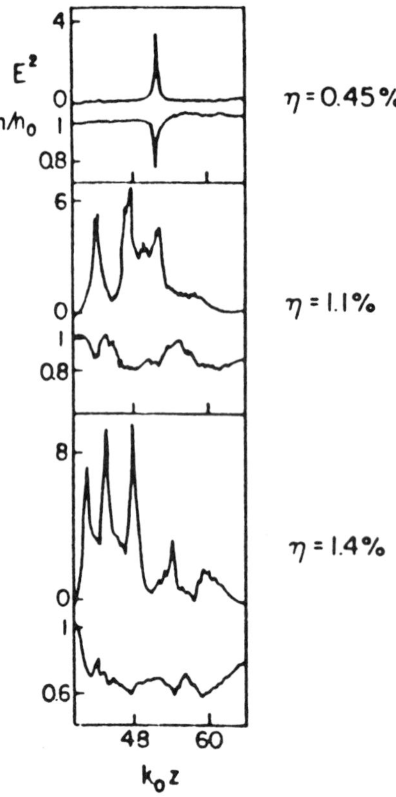

Figure 5.12. Dependence of the electric field spatial structures on the beam density. (Due to Cheung and Wong [75], by courtesy of The American Institute of Physics.)

Rowlands [81], and Shivamoggi [82]. Shimizu and Ichikawa [78] used the reductive perturbation method; Kakutani and Sugimoto [79] and Chan and Seshadri [80] used the Krylov—Bogliubov—Mitropolski method to derive a nonlinear Schrödinger equation which predicted the ion-acoustic waves to be stable to modulations of long-wavelength. Shivamoggi [82] derived a different nonlinear Schrödinger equation for the ion-acoustic waves instead by using the Zakharov—Karpman approach wherein the ion-acoustic wave interacts with a slow plasma motion, and obtained the same result.

Consider one-dimensional ion-acoustic waves travelling in a plasma consisting of cold ions and hot and isothermal electrons. Neglecting the electron inertia (since $m_e/m_i \ll 1$), one has the following governing equations,

$$\frac{\partial n_i}{\partial t} + \frac{\partial}{\partial x}(n_i V_i) = 0 \tag{148}$$

$$\frac{\partial V_i}{\partial t} + V_i \frac{\partial V_i}{\partial x} = \frac{e}{m_i} E \tag{149}$$

$$\frac{\partial n_e}{\partial x} = -\frac{en_e}{KT_e} E \tag{150}$$

$$\frac{\partial E}{\partial x} = 4\pi e(n_i - n_e). \tag{151}$$

One may derive from equations (148)—(151), for a finite-amplitude ion-acoustic wave

$$\frac{\partial}{\partial t}\left[\left(1+\frac{\delta n_e}{n_0}\right)\frac{\partial V_i}{\partial t}\right] - \lambda_D^2 \frac{\partial^4 V_i}{\partial t^2 \partial x^2}$$

$$- C_s^2 \frac{\partial^2}{\partial x^2}\left[\left(1+\frac{\delta n_i}{n_0}\right)V_i\right] = 0 \tag{152}$$

where

$$\lambda_D^2 = \frac{V_{T_e}^2}{\omega_{p_e}^2}, \quad C_s^2 = \frac{KT_e}{m_i}$$

and δn_e, δn_i are the low-frequency perturbations in the number densities of the electrons and ions, respectively, associated with a slow plasma motion.

Let us look for solutions of the form

$$V_i(x,t) = \tilde{V}_i(x,t)e^{-i\omega_0 t} \tag{153}$$

where $\tilde{V}_i(x, t)$ is a slowly-varying function of t. Let us consider only long-wavelength perturbations, and assume $\delta n_e \approx \delta n_i$. Then, noting that $\omega_0 \approx kC_s$, (152) gives

$$\frac{2i}{\omega_0}\frac{\partial \tilde{V}_i}{\partial t} - \lambda_D^2 \frac{\partial^2 \tilde{V}_i}{\partial x^2} - \frac{k^2\lambda_D^2}{1+k^2\lambda_D^2}\left(\frac{\delta n_e}{n_0}\right)\tilde{V}_i = 0. \tag{154}$$

Let us decompose the electric field E into a high-frequency part $\tilde{E}(x, t)$ $e^{-i\omega_0 t}$ (where $\tilde{E}(x, t)$ varies slowly in time) and a low-frequency part $-\partial\phi/\partial x$. In view of the fact that the phase velocity of the ion-acoustic waves is much less than the thermal velocity of the electrons, we may describe the low-frequency motion of the electrons by means of the Boltzmann description:

$$\frac{\delta n_e}{n_0} = \exp - \left[\left(-e\phi + \frac{e^2|E|^2}{4m_e\omega_0^2}\right)\frac{1}{KT_e}\right] - 1$$

$$\approx \left(e\phi - \frac{e^2|E|^2}{4m_e\omega_0^2}\right)\frac{1}{KT_e} \tag{155}$$

where the second term in the exponent represents the ponderomotive force produced by the high-frequency part of the electric field on the slow-time scale.

The equations governing linearized low-frequency perturbations for the ions are

$$\frac{\partial V_i}{\partial t} = -\frac{e}{m_i}\frac{\partial \phi}{\partial x} \tag{156}$$

$$\frac{\partial}{\partial t}\delta n_i + n_0 \frac{\partial V_i}{\partial x} = 0. \tag{157}$$

For a stationary propagation with velocity V, the various quantities depend on x and t only in the combination $(x - Vt)$ so that equations (155)–(157) give, on recalling $\delta n_e \approx \delta n_i$,

$$\frac{\delta n_e}{n_0} = \frac{e^2 |\tilde{E}|^2/4m_i\omega_0^2}{V^2 - C_s^2}. \tag{158}$$

Using (158), and noting from equation (159) that, to first approximation,

$$\tilde{V}_i \approx \frac{e\tilde{E}}{m_i\omega_0} \tag{159}$$

equation (144) gives the nonlinear Schrödinger equation

$$2i\omega_0 \frac{\partial \tilde{E}}{\partial t} - \omega_0^2\lambda_D^2 \frac{\partial^2 \tilde{E}}{\partial x^2}$$

$$- \left[\frac{k^4\lambda_D^2 C_s^2}{1 + k^2\lambda_D^2}\frac{e^2/4m_i\omega_0^2}{V^2 - C_s^2}\right] |\tilde{E}|^2\tilde{E} = 0. \tag{160}$$

Now since the modulation is assumed to be slow compared with $1/\omega_0$, one requires $V \ll C_s$. This means that in equation (160), one has

$$(-\omega_0^2\lambda_D^2)\left[\frac{k^4\lambda_D^2 C_s^2}{1 + k^2\lambda_D^2}\frac{(-e^2/4m_i\omega_0^2)}{V^2 - C_s^2}\right] < 0 \tag{161}$$

which implies in turn that the ion-acoustic waves are stable to slow modulations of long wavelength.*

* This result can also be deduced alternately as follows: Let us go back to original Korteweg–de Vries equation that governs the nonlinear ion-acoustic waves

$$\frac{\partial \phi}{\partial t} + \phi\frac{\partial \phi}{\partial x} + \frac{1}{2}\frac{\partial^3 \phi}{\partial x^3} = 0$$

and look for solutions of the form:

$$\phi = \sum_{n=0}^{\infty} \varepsilon^n \sum_{m=-\infty}^{\infty} \phi_m^{(n)}(\xi, \tau) e^{im(kx - \omega t)}$$

where $\varepsilon \ll 1$ is a small parameter and

$$\xi = \varepsilon(x - \sigma t),\ \tau = \varepsilon^2 t.$$

The nonlinear modulation of the ion wave packet was observed experimentally by Watanabe [184]. The modulation was observed in both cases where the carrier was composed of broad-band frequency and where the carrier frequency was quasi-monochromatic. In both cases, the wave packet was modulationally unstable when the amplitude was large. This threshold is not predicted by the nonlinear Schrödinger equation because the latter does not include dissipation due to Landau damping. Ikezi *et al.* [83] made an experiment on the nonlinear evolution of an ion-acoustic wave packet and found that the self-modulation together with dispersive effects causes the wave packet to expand and to break into two packets. The ion-acoustic wave was stable with respect to the modulational perturbations. However, modulational instability did not appear to have any chance to occur because of the dominant effects introduced by the ions trapped in the wave potential troughs. Honzawa and Hollenstein [185] performed experiments on the nonlinear self-modulation of the fast ion beam mode. In their ion beam-plasma system with beam velocity $V_b \approx C_s$, only the fast beam mode was observed to propagate at distances far from the excitation plane. At large amplitudes above the threshold ($\delta n/n \sim 0.01$) the beam mode (carrier wave) was observed to be self amplitude-modulated. Figure 5.13 shows the occurrence of the amplitude modulation at large amplitudes above threshold and at large distances from the grid. Observe that the instability appearing on the wave envelope grows rapidly with increasing distance and was then saturated. When the self-modulation had fully evolved, the carrier wave was observed to split into a train of wave packets (Figure 5.14).

The nonlinear development of ion-acoustic waves in a magnetized plasma was investigated by Zakharov and Kuznetsov [186], Laedke and Spatschek [187, 188], Infeld [189], and Shivamoggi [190]. Here the dispersion arising from charge separation as well as finite-gyroradius effects can balance the nonlinearity.

We then obtain:

$$0(\varepsilon):\quad m = \pm 1:\quad \omega = -\frac{1}{2}k^3$$

$$m \neq \pm 1:\quad \phi_m^{(1)} = 0$$

$$0(\varepsilon^2):\quad m = \pm 1:\quad \sigma = -\frac{3}{2}k^2$$

$$m = 2:\quad \phi_2^{(2)} = \frac{1}{3k^2}[\phi_1^{(1)}]^2$$

$$0(\varepsilon^3):\quad m = 0:\quad \phi_0^{(2)} = \frac{1}{\sigma}|\phi_1^{(1)}|^2$$

$$m = 1:\quad i\frac{\partial \phi_1^{(1)}}{\partial \tau} - \frac{3}{2}k\frac{\partial^2 \phi_1^{(1)}}{\partial \xi^2} + \frac{1}{3k}|\phi_1^{(1)}|^2\phi_1^{(1)} = 0.$$

The above nonlinear Schrödinger equation immediately establishes that the ion-acoustic waves are modulationally stable.

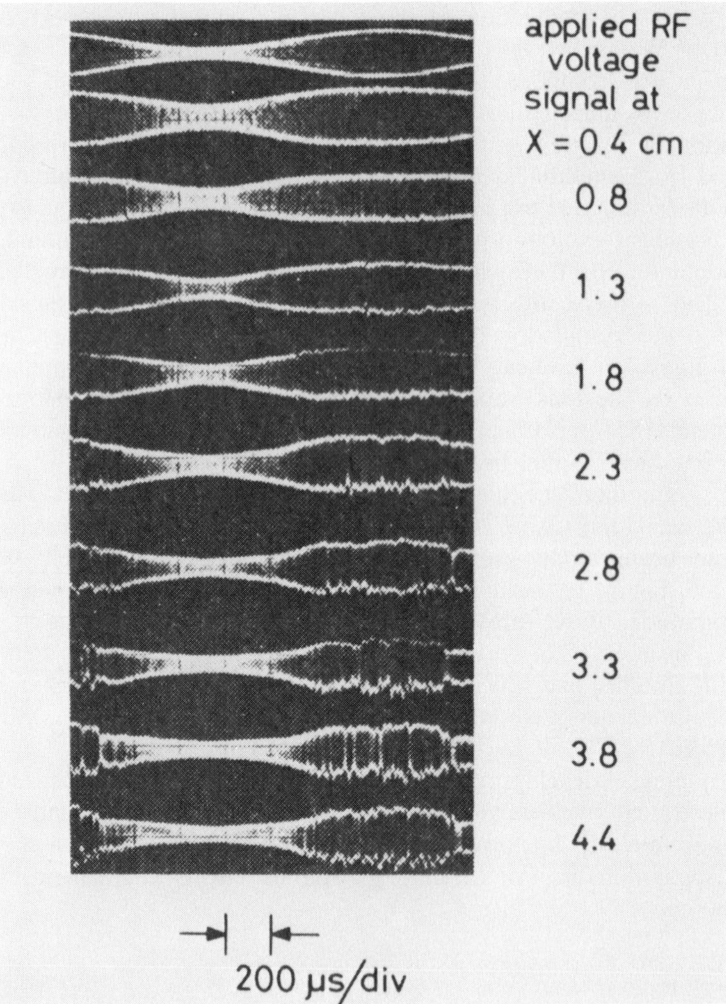

Figure 5.13. Carrier wave signals evolving with distance x. These are obtained at $\omega_0/2\pi$ = 587 kHz. Here, the applied rf voltage (top trace) is externally modulated at 620 Hz. (Due to Honzawa and Hollenstein [185], by courtesy of The American Institute of Physics.)

(iii) Upper-Hybrid Solitons

Modulational instability and formation of envelope solitons of the upper hybrid wave have been considered by Kaufman and Stenflo [84], Porkolab and Goldman [85], and Shivamoggi [62, 86]. It turns out that a strong magnetic field can eliminate the existence of subsonic upper-hybrid solitons but can make possible the existence of supersonic upper-hybrid solitons. In order to treat the upper-hybrid solitons travelling at speeds near the speed of sound, the ion-nonlinearity has to be taken into account (Shivamoggi [62]).

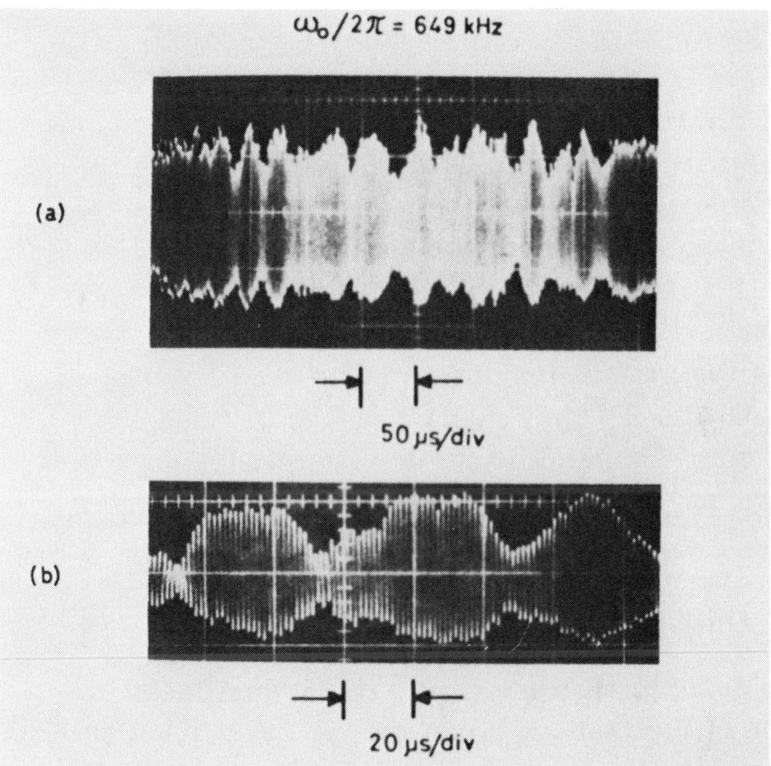

Figure 5.14. Shapes of the wave envelopes. (a) Typical carrier wave envelope splitting into a train of wave packets. The individual wave packets are somewhat irregular in shape. (b) Oscilloscope trace of the carrier wave signal forming a few wave packets. (Due to Honzawa and Hollenstein [185], by courtesy of The American Institute of Physics.)

(a) Subsonic and Supersonic Upper-Hybrid Solitons

Consider a plasma subjected to a magnetic field **B** directed along the z-direction. The dispersion relation for the upper-hybrid waves in the limit $k^2KT_e/m_e\Omega_e^2 \ll 1$ (here k is the wavenumber, and $\Omega_e = eB/m_ec$) is given by (Krall and Trivelpiece [119])

$$\omega^2 = \omega_u^2 + \frac{3k^2V_{T_e}^2}{1 - 3\Omega_e^2/\omega_{p_e}^2} \tag{162}$$

where,

$$\omega_u^2 = \omega_{p_e}^2 + \Omega_e^2.$$

Let us take the electric field associated with the upper-hybrid wave to be $\mathbf{E} = E\hat{\mathbf{i}}_x$ (the z-component of E can be taken to be small and negligible). Converting

(162) into an operator equation, one obtains

$$\frac{\partial^2 E}{\partial t^2} + \omega_u^2 E - \frac{3V_{T_e}^2}{(1 - 3\Omega_e^2/\omega_{p_e}^2)} \frac{\partial^2 E}{\partial x^2} = 0. \tag{163}$$

Putting,

$$E(x, t) = \tilde{E}(x, t)\, e^{-i\omega_u t} \tag{164}$$

$\tilde{E}(x, t)$ being a slowly-varying function of t, equation (163) gives

$$2i\omega_u \frac{\partial \tilde{E}}{\partial t} + 3V_{T_e}^2 \left(1 - \frac{3\Omega_e^2}{\omega_{p_e}^2}\right) \frac{\partial^2 \tilde{E}}{\partial x^2} = (\delta n_e)\,\omega_{p_e}^2 \tilde{E} \tag{165}$$

where δn_e is the low-frequency electron density perturbation.

The potential associated with the ponderomotive force exerted by the upper-hybrid waves on the electrons is given by averaging the nonlinear term $(m_e \mathbf{V}_e \cdot \nabla \mathbf{V}_e)_x$ over period of $0(\omega_u^{-1})$. Thus, noting that for a single electron

$$V_{e_x} = \frac{ie}{m_e} \frac{\omega_u}{\omega_u^2 - \Omega_e^2} \tilde{E}_x, \quad V_{e_y} = \frac{e}{m_e} \frac{\Omega_e}{\omega_u^2 - \Omega_e^2} E_x,$$
$$V_{e_z} = \frac{ie}{m_e \omega_u} \tilde{E}_z \tag{166}$$

one may calculate for the ponderomotive force potential:

$$\psi_e = \frac{e}{4} \frac{|\tilde{E}|^2}{\omega_{p_e}^2}. \tag{167}$$

Upon averaging the low-frequency electron motion parallel to the magnetic field over periods of $0(\omega_u^{-1})$, one obtains

$$-\frac{KT_e}{n_0} \frac{\partial}{\partial z}(\delta n_e) + e\frac{\partial \phi}{\partial z} - \frac{e^2}{4m_e\omega_{p_e}^2} \frac{\partial}{\partial z} |\tilde{E}|^2 = 0 \tag{168}$$

where ϕ is the low-frequency ambipolar potential.

One has for the ion motion, noting that the ponderomotive force is smaller than that on the electrons by a factor of m_e/m_i,

$$\frac{\partial}{\partial t}(\delta n_i) + n_0 \frac{\partial V_{i_x}}{\partial x} = 0 \tag{169}$$

$$\frac{\partial V_{i_x}}{\partial t} = -\frac{e}{m_i} \frac{\partial \phi}{\partial x} + \Omega_i V_{i_y} \tag{170}$$

$$\frac{\partial V_{i_y}}{\partial t} = -\Omega_i V_{i_x} \tag{171}$$

where δn_i is the ion-density perturbation. One obtains from equations (169)–(171):

$$\frac{\partial^2}{\partial t^2}(\delta n_i) + \Omega_i^2 \delta n_i - \frac{e n_0}{m_i}\frac{\partial^2 \phi}{\partial x^2} = 0. \tag{172}$$

Assuming charge neutrality in the low-frequency motion, i.e., $\delta n_e \approx \delta n_i = \delta n$, equations (168) and (172) give the ion-cyclotron wave

$$\left(\frac{\partial^2}{\partial t^2} - C_s^2 \frac{\partial^2}{\partial x^2} + \Omega_i^2\right)\delta n_e = \frac{\partial^2}{\partial x^2}\left(\frac{|\tilde{E}|^2}{16\pi m_i}\right). \tag{173}$$

Let us now nondimensionalize t, x, δn and $|E|$ by ω_{pi}^{-1}, $(KTe/m_e\omega_{p_e}^2)^{1/2}$, n_0, and $2\sqrt{\pi n_0 T_e}$. Let us next assume that $kC_s \gg \Omega_i$. (It is easy to verify that this assumption is compatible with the condition $k^2 KT_e/m_e\Omega_e^2 \ll 1$ we have already imposed on the solution.) Then equations (165) and (163) become

$$i\varepsilon \tilde{E}_t + \frac{3/2}{(1-3\alpha^2)(1+\alpha^2)^{1/2}}\tilde{E}_{xx} = \frac{\delta n}{2(1+\alpha^2)^{1/2}}\tilde{E} \tag{174}$$

$$\left(\frac{\partial^2}{\partial t^2} - \frac{\partial^2}{\partial x^2}\right)\delta n = \frac{\partial^2}{\partial x^2}\left(\frac{|\tilde{E}|^2}{4}\right) \tag{175}$$

where,

$$\varepsilon \equiv \sqrt{\frac{m_e}{m_i}}, \quad \alpha \equiv \frac{\Omega_e}{\omega_{p_e}}.$$

Let us now look for stationary waves

$$\delta n(x, t) = \delta n(x - Mt) \tag{176}$$

so that equation (175) gives:

$$\delta n = -\frac{|\tilde{E}|^2}{4(1-M^2)}. \tag{177}$$

Using (177), equation (175) gives

$$i\varepsilon \tilde{E}_t + \frac{3/2}{(1-3\alpha^2)(1+\alpha^2)^{1/2}}\tilde{E}_{xx} + \\ + \frac{1/8}{(1-M^2)(1+\alpha^2)^{1/2}}|\tilde{E}|^2\tilde{E} = 0. \tag{178}$$

It is obvious from equation (178) that the modulational instability of upper-hybrid waves arises if

$$(1-3\alpha^2)(1-M^2) > 0. \tag{179}$$

Therefore, accordingly as $\alpha^2 \lessgtr 1/3$, one may have subsonic or supersonic envelope solitons of the upper-hybrid waves. In order to determine the latter explicitly, put

$$\tilde{E}(x,t) = \mathscr{E}(x - Mt)\, e^{i(\varepsilon\mu x - \sigma t)} \tag{180}$$

where,

$$\mu = \frac{M}{3}(1+\alpha^2)^{1/2}(1-3\alpha^2)$$

so that equation (178) gives

$$\mathscr{E}'' - \frac{3}{2\beta}\left(\frac{\varepsilon^2 M^2}{6\beta} - \varepsilon\sigma\right)\mathscr{E} + \frac{\nu}{12}\mathscr{E}^3 = 0 \tag{181}$$

where,

$$\beta \equiv \frac{1}{(1+\alpha^2)^{1/2}(1-3\alpha^2)}, \quad \nu \equiv \frac{1-3\alpha^2}{1-M^2}.$$

Thus,

$$\mathscr{E} = \left[\frac{4}{3}(1-M^2)(1+\alpha^2)\left\{\frac{\varepsilon^2 M^2}{6}(1-3\alpha^2) - \frac{\varepsilon\sigma}{(1+\alpha^2)^{1/2}}\right\}\right] \times$$
$$\times \operatorname{sech}\left[\frac{2}{3}(1-3\alpha^2)(1+\alpha^2)^{1/2} \times\right.$$
$$\left.\times\left\{\frac{\varepsilon^2 M^2}{6}(1-3\alpha^2)(1+\alpha^2)^{1/2} - \varepsilon\sigma\right\}\right]^{1/2}(x - Mt). \tag{182}$$

Observe that although the solitons may exist for $M \lessgtr 1$ according as $\alpha^2 \lessgtr 1/3$ they disappear as $M \Rightarrow 1$. In order to obtain a correct representation for the latter case, one has to include the ion-nonlinearity in the low-frequency response of the plasma to the upper-hybrid waves (Shivamoggi [62]).

Experimental confirmation of the modulational instability at the upper-hybrid frequency and the formation of an upper-hybrid soliton was made by Cho and Tanaka [87]. They injected a high powered microwave p_μ in the form of the extraordinary mode into an afterglow plasma column in a uniform magnetic field. The incident microwave, passing the cyclotron cutoff, tunneled through the evanescent region and arrived at the upper-hybrid resonance layer ($\omega = \omega_u$), where it was mode-converted to the electron Bernstein mode (EBM) which propagated towards the high-density region. Figure 5.15a shows the radial profiles on the wave intensity I_μ and the electron density n_e when the low p_μ was injected from the left-hand side of the plasma column. However, when a high p_μ was injected, the wave intensity was enhanced strongly at the upper-hybrid resonance layer, and its ponderomotive force led to the formation of the density cavity there. The injected microwave seemed to be trapped at the upper-hybrid resonance layer, since the intensity I_μ at $\omega = \omega_u$ increased

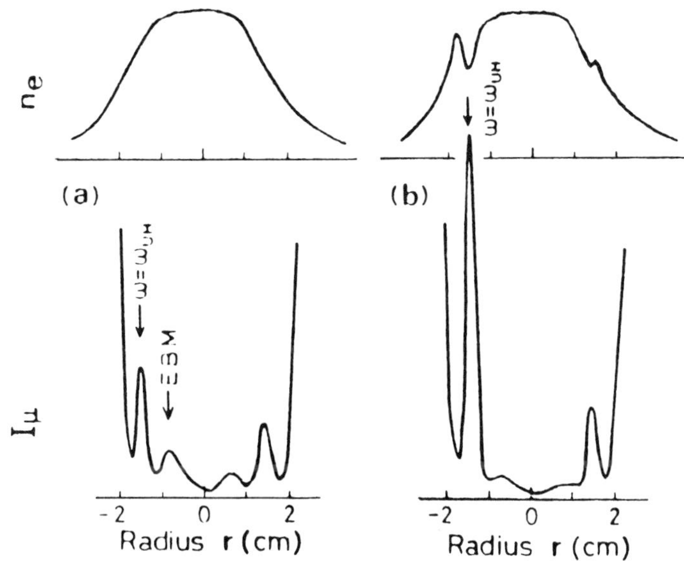

Figure 5.15. Radial profiles of the wave intensity I_μ and the electron density n_e (electron saturation current I_e measured with a Langmuir probe). (a) Incident microwave power P_μ = 0.5 kW, (b) 30 kW (τ_μ = 350 ns), respectively. The intensity I_μ is detected at t = 200 ns after P_μ is turned on, while the current I_e is sampled at t = 150 ns after P_μ is turned off. In (a) and (b) the microwave attenuators in the input and the detecting circuits are adjusted to receive the same level of I_μ in the absence of the plasma. B = 0.6 kG, n_{e0} = 1.5 × 10^{12} cm^{-3} and p = 5 × 10^{-4} Torr. (Due to Cho and Tanaka [87], by courtesy of The American Physical Society.)

drastically, while that of the electron Bernstein mode decreased, compared with the case of weak p_μ (Figure 5.15b).

(b) Sonic Upper-Hybrid Solitons

For the low-frequency waves associated with the ion motions, the electrons will remain in thermo-dynamic equilibrium at a constant temperature T_e with a number density given by the Boltzmann distribution

$$n_e = n_0 \exp\left(\frac{e\phi}{KT_e} - \frac{|\tilde{E}|^2}{16\pi n_0 KT_e}\right). \tag{183}$$

One then has for the ion motions

$$\frac{\partial}{\partial t}(\delta n_i) + n_0 \frac{\partial V_{i_x}}{\partial x} = 0 \tag{184}$$

$$m_i \frac{\partial V_{i_x}}{\partial t} = -e \frac{\partial \phi}{\partial x} + \Omega_i V_{i_y} \tag{185}$$

$$m_i \frac{\partial V_{i_y}}{\partial t} = -\Omega_i V_{i_x} \tag{186}$$

$$\frac{\partial^2 \phi}{\partial x^2} = -4\pi \left[\delta n_i - n_0 \left\{ \exp\left(\frac{e\phi}{KT_e} - \frac{|\tilde{E}|^2}{16\pi n_0 KT_e} \right) - 1 \right\} \right]. \tag{187}$$

Nondimensionalizing as before, and assuming $kC_s \gg \Omega_1$, equations (184)–(187) give

$$\left(\frac{\partial^2}{\partial t^2} - \frac{\partial^2}{\partial x^2} \right) \delta n_i - \frac{\partial^4}{\partial x^2 \partial t^2} \delta n_i + \frac{\partial^2}{\partial x^2} (\delta n_i)^2$$

$$= \frac{\partial^2}{\partial x^2} \left(\frac{|\tilde{E}|^2}{4} \right). \tag{188}$$

Let us rescale x and $\tilde{E}$ so that

$$x' = (1 - 3\alpha^2)^{1/2} x, \quad \tilde{E}' = \tilde{E}(1 + \alpha^2)^{-1/2} \tag{189}$$

and drop the primes. Then, equations (184) and (188) become

$$i\varepsilon(1 + \alpha^2)\tilde{E}_t + \frac{3}{2} \tilde{E}_{xx} = \frac{\delta n_e}{2} \tilde{E} \tag{190}$$

$$\left(\frac{1}{1 - 3\alpha^2} \frac{\partial^2}{\partial t^2} - \frac{\partial^2}{\partial x^2} \right) \delta n_i - \frac{\partial^4}{\partial x^2 \partial t^2} \delta n_i$$

$$+ \frac{\partial^2}{\partial x^2} (\delta n_i)^2 = \frac{\partial^2}{\partial x^2} \left[(1 + \alpha^2) \frac{|\tilde{E}|^2}{4} \right]. \tag{191}$$

Let us assume $\delta n_e \approx \delta n_i = \delta n$, and look for a solution of the form

$$\tilde{E}(x, t) = \mathscr{E}(\xi) \exp\left(i \frac{\varepsilon M}{3} \sqrt{1 + \alpha^2} x - i\sigma t \right) \tag{192}$$

$$\delta n(x, t) = \delta n(\xi), \quad \xi \equiv x - Mt$$

so that equations (190) and (191) give

$$\mathscr{E}'' - \frac{2}{3} \left[\frac{\varepsilon^2 M^2}{6} (1 + \alpha^2) - \varepsilon\sigma \right] \mathscr{E} - \frac{\delta n}{3} \mathscr{E} = 0 \tag{193}$$

$$\left(\frac{M^2}{1 - 3\alpha^2} - 1 \right) \delta n - M^2 \delta n'' + \delta n^2 = \frac{(1 + \alpha^2)}{4} \mathscr{E}^2. \tag{194}$$

Let us look for a solution of the form

$$\begin{aligned} \mathscr{E} &= A \operatorname{sech} \xi \\ \delta n &= B \operatorname{sech}^2 \xi \end{aligned} \tag{195}$$

so that equations (193) and (194) give

$$\left[-\gamma^2 - \frac{B}{3} - \frac{2}{3}\left\{\frac{\varepsilon^2 M^2}{6}(1+\alpha^2) - \varepsilon\sigma\right\}\right] + \left(2\gamma^2 + \frac{B}{3}\right)\tanh^2 \gamma\xi = 0 \tag{196}$$

$$\left[\left(\frac{M^2}{1-3\alpha^2} - 1\right)B + 2M^2\gamma^2 B + B^2 - \frac{(1+\alpha^2)}{4}A^2\right] + \left[-\left(\frac{M^2}{1-3\alpha^2} - 1\right)B - 8M^2\gamma^2 B - 2B^2 + \frac{(1+\alpha^2)}{4}A^2\right]\tanh^2 \gamma\xi + (6M^2\gamma^2 B + B^2)\tanh^4 \gamma\xi = 0. \tag{197}$$

Comparison of the coefficients of the last term in (196) and (197) shows that the assumed form of the solution in (195) is valid only for $M \approx 1$. One obtains from (196) and (197)

$$\begin{aligned} B &= -\frac{4M^2\left[\dfrac{\varepsilon M^2}{6}(1+\alpha^2) - \varepsilon\sigma\right]}{2M^2 - 1} \\ A^2 &= \frac{4}{1+\alpha^2}\left[\left(\frac{M^2}{1-3\alpha^2} - 1\right)B + \frac{2M^2 B}{3}\right], \\ \gamma^2 &= -\frac{B}{6M^2}. \end{aligned} \tag{198}$$

Observe that this solution is well behaved for $M = 1$.

(c) Upper-Hybrid Solitons in an Electron Plasma

We now consider upper-hybrid waves in an electron-plasma coexisting with a uniform immobile positive ion background. We will establish the existence of upper-hybrid solitons in the electron-plasma by explicitly evaluating the amplitude-dependent frequency shift for the upper hybrid wave produced by the nonlinear effects in a cold plasma and by introducing then the effects due to thermal dispersion in the warm plasma (Shivamoggi [86]). Since the ion motions are neglected, the Zakharov mechanism for the production of upper-hybrid solitons (considered in the foregoing) would predict the latter to exist only at supersonic speeds or when the thermal dispersion is negative. The latter is indeed found to be the condition for the existence of upper-hybrid solitons in an electron-plasma!

Consider a cold electron-plasma subjected to a uniform magnetic field $\mathbf{B} = B\hat{\mathbf{i}}_z$. The one-dimensional electrostatic wave motions occurring perpendicular to the applied magnetic field are governed by the following equations

$$\frac{\partial n}{\partial t} + \frac{\partial}{\partial x}(nV_x) = 0 \tag{199}$$

$$\frac{\partial V_x}{\partial t} + V_x \frac{\partial V_x}{\partial x} = \frac{e}{m_e}\frac{\partial \phi}{\partial x} - \Omega_e V_y \tag{200}$$

$$\frac{\partial V_y}{\partial t} + V_x \frac{\partial V_y}{\partial x} = \Omega_e V_x \tag{201}$$

$$\frac{\partial^2 \phi}{\partial x^2} = 4\pi e n. \tag{202}$$

For travelling waves, the various quantities depend on x and t, only in the combination $\xi = kx - \omega t$. Then, one may derive the equations (199)–(202):

$$(\Omega_e^2 + \omega_{p_e}^2)\frac{\mathrm{d}\phi}{\mathrm{d}\xi}\left(1 + \frac{k^2}{en_0}\frac{\mathrm{d}^2\phi}{\mathrm{d}\xi^2}\right)^3 + \\ + \omega^2 \frac{\mathrm{d}^3\phi}{\mathrm{d}\xi^3}\left(1 + \frac{2k^2}{en_0}\frac{\mathrm{d}^2\phi}{\mathrm{d}\xi^2}\right) = 0. \tag{203}$$

Let us introduce a small parameter $\varepsilon \ll 1$ which may characterize the magnitude of a typical perturbation, and seek solutions to equation (193) of the form

$$\phi(\xi;\varepsilon) = \sum_{n=1}^{\infty} \varepsilon^n \phi_n(\xi) \\ \omega(k;\varepsilon) = \sum_{n=0}^{\infty} \varepsilon^n \omega_n(k). \tag{204}$$

We are using the method of strained parameters.

Using (204), equation (203) gives

$$(\Omega_e^2 + \omega_{p_e}^2)\frac{d\phi_1}{d\xi} + \omega_0^2\frac{d^3\phi_1}{d\xi^3} = 0 \tag{205}$$

$$\begin{aligned}(\Omega_e^2 + \omega_{p_e}^2)\frac{d\phi_2}{d\xi} + \omega_0^2\frac{d^3\phi_2}{d\xi^3} \\ = -3(\Omega_e^2 + \omega_{p_e}^2)\frac{k^2}{en_0}\frac{d\phi_1}{d\xi}\frac{d^2\phi_1}{d\xi^2} - \\ - 2\omega_0^2\frac{k^2}{en_0}\frac{d^2\phi_1}{d\xi^2}\frac{d^3\phi_1}{d\xi^3} - 2\omega_0\omega_1\frac{d^3\phi_1}{d\xi^3}\end{aligned} \tag{206}$$

$$\begin{aligned}(\Omega_e^2 + \omega_{p_e}^2)\frac{d\phi_3}{d\xi} + \omega_0^2\frac{d^3\phi_3}{d\xi^3} = -3(\Omega_e^2 + \omega_{p_e}^2)\left[\frac{k^2}{en_0}\frac{d\phi_2}{d\xi}\frac{d^2\phi_1}{d\xi^2} + \right. \\ \left. + \frac{k^4}{e^2n_0^2}\frac{d\phi_1}{d\xi}\left(\frac{d^2\phi_1}{d\xi^2}\right)^2 + \frac{k^2}{en_0}\frac{d\phi_1}{d\xi}\frac{d^2\phi_2}{d\xi^2}\right] - \\ - 2\omega_0^2\frac{k^2}{en_0}\left[\frac{d^3\phi_1}{d\xi^3}\frac{d^2\phi_2}{d\xi^2} + \frac{d^3\phi_2}{d\xi^3}\frac{d^2\phi_1}{d\xi^2}\right] - \\ - (2\omega_0\omega_2 + \omega_1^2)\frac{d^3\phi_1}{d\xi^3}.\end{aligned} \tag{207}$$

One obtains from equation (205), the familiar linear result

$$\begin{aligned}\phi_1 &= A\cos\xi \\ \omega_0^2 &= \omega_{p_e}^2 + \Omega_e^2.\end{aligned} \tag{208}$$

Using (208), the removal of secular terms on the right hand side of equation (206) requires

$$\omega_1 \equiv 0 \tag{209}$$

and then the solution to equation (206) is given by

$$\phi_2 = \frac{A^2k^2}{12en_0}\sin 2\xi. \tag{210}$$

Using equations (208)—(210), the condition for the removal of secular terms on the right hand side of equation (207) gives

$$\omega_2 = \frac{11}{24}\frac{k^4}{e^2n_0^2}(\Omega_e^2 + \omega_{p_e}^2)^{1/2}A^2. \tag{211}$$

Thus, the frequency ω is given by

$$\omega = (\omega_{p_e}^2 + \Omega_e^2)^{1/2}\left[1 + \varepsilon^2 \frac{11}{24}\frac{k^4}{e^2 n_0^2}|\phi_1|^2 + \cdots\right]. \tag{212}$$

Equation (212) shows that purely-periodic upper-hybrid waves propagating perpendicular to the applied magnetic field exist even when their amplitude becomes finite, with the effects of the latter showing up as an amplitude-dependent shift in the frequency. A similar result was deduced by Tidman and Stainer [186] for cyclotron waves propagating along the applied magnetic field.

Let us now consider the plasma to be warm, and assume that the thermal dispersion to be weak. Then, one may write in place of (212),

$$\omega^2 = \omega_u^2 + \varepsilon^2 \frac{11}{12}\frac{k^4}{e^2 n_0^2}\omega_u^2|\phi_1|^2 + \frac{3k^2 V_{T_e}^2}{(1 - 3\Omega_e^2/\omega_{p_e}^2)} \tag{213}$$

where

$$\omega_u^2 \equiv \omega_{p_e}^2 + \Omega_e^2.$$

Equation (213) shows that the modulational instability of the upper-hybrid waves arises if

$$\left(\varepsilon^2 \frac{11}{12}\frac{k^4}{e^2 n_0^2}\right)\left(\frac{3k^2 V_{T_e}^2}{1 - 3\Omega_e^2/\omega_{p_e}^2}\right) < 0 \tag{214}$$

or

$$\omega_{p_e}^2 < 3\Omega_e^2$$

i.e., when the thermal dispersion is negative. This result is consistent with the existence of supersonic upper-hybrid solitons associated with negative thermal dispersion in Zakharov's model (considered before) for which the ion motions can indeed be neglected!

(iv) Lower-Hybrid Solitons

Gekelman and Stenzel [88] performed an experiment on the self-focusing of large-amplitude lower-hybrid waves. Wave bursts at $\omega_0 > \omega_{LH}$ (ω_{LH} being the lower-hybrid frequency) were excited from a grid in a quiescent magnetized plasma column and propagated into a density gradient perpendicular to magnetic field $\mathbf{B}_0$. The subsequent evolution of the internal rf field and the plasma parameters was investigated. As the amplitude of the applied burst was increased, two nonlinear effects were observed: the internal rf fields became localized in a spatial wave packet propagating into the plasma along a conical trajectory which made a small angle with respect to the confining magnetic field and, simultaneously, strong density depressions were formed at the region of the highly-peaked rf field. Figure 5.16 gives the large-amplitude distribution inside

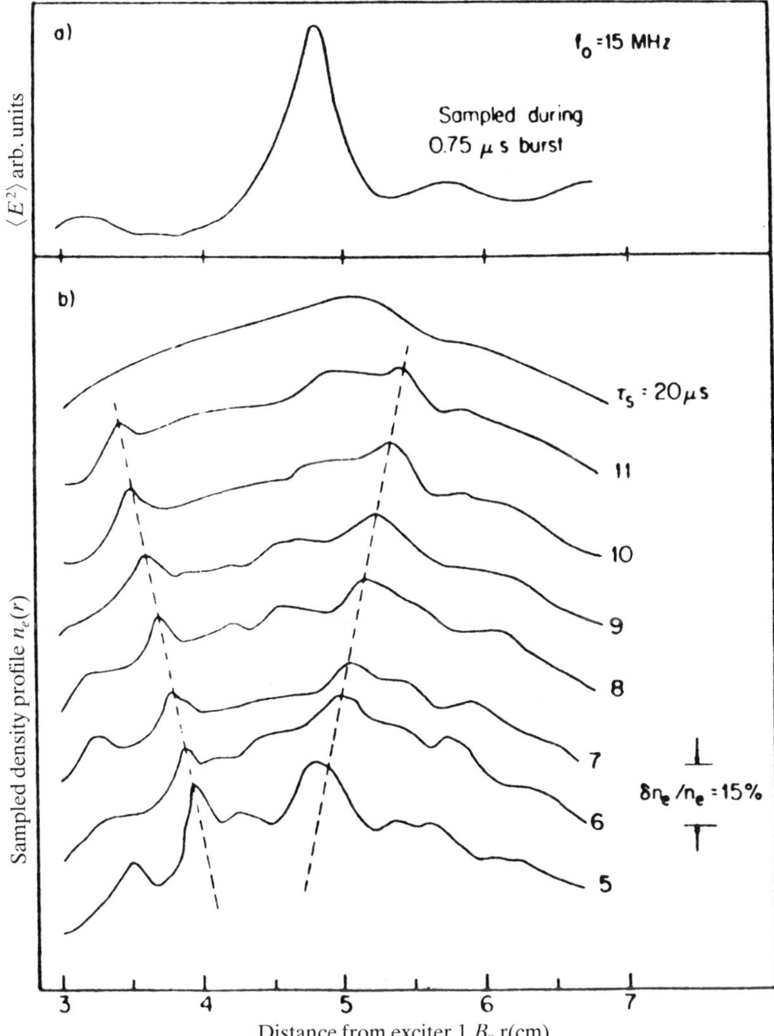

Figure 5.16. (a) Time-averaged square of the electric field versus radial position in the nonlinear case, sampled during the final 250 nsec of the 750-nsec burst. Plasma density, B_0, and repetition rate are the same as in Fig. 1. (b) Sampled radial density profile (zero lines suppressed) at various times τ_s after the end of the exciter burst. The lowest curve (τ_s = 5μsec) indicates a density perturbation of 15%. (Due to Gekelman and Stenzel [88], by courtesy of The American Physical Society.)

the column (top trace) and the modified density profile at different times after the end of the rf burst (bottom traces). The density perturbation and the rf field peaks are inclined to the magnetic field at approximately the resonance cone angle. Morales and Lee [89] traced these conical wave packets to the filamentation through self-interaction of a large-amplitude lower-hybrid wave in the plasma. This was suggested to occur through the Zakharov mechanism wherein density cavities are created by the ponderomotive force exerted by the lower-hybrid waves on both the electrons and ions which in turn trap the latter. We have,

$$\frac{\partial n_s}{\partial t} + \nabla \cdot (n_s \mathbf{V}_s) = 0 \tag{215}$$

$$\frac{\partial \mathbf{V}_s}{\partial t} + (\mathbf{V}_s \cdot \nabla)\mathbf{V}_s = \frac{q_s}{m_s}\left(-\nabla\phi + \frac{1}{c}\mathbf{V}_s \times \mathbf{B}\right) - \frac{V_{T_s}^2}{n_s}\nabla n_s \tag{216}$$

$$-\frac{\partial}{\partial t}\nabla^2\phi = 4\pi\nabla \cdot \sum_s q_s n_s \mathbf{V}_s \tag{217}$$

where the subscript $s = e, i$ refers to the electrons and ions, respectively, q is the electric charge, and $\mathbf{E} = -\nabla\phi$ is the lower-hybrid electric field.

Let $\mathbf{B} = B\hat{\mathbf{i}}_z$ and the ions be cold and electrons be hot, i.e., $T_e \gg T_i$. Assuming,

$$\phi(\mathbf{x}, t) = \phi(\mathbf{x})e^{i\omega t} \tag{218}$$

equations (215)—(217) give

$$\nabla^2\phi + i\frac{4\pi e}{\omega}\nabla \cdot (n_e\boldsymbol{\mu}_e - n_i\boldsymbol{\mu}_i) \cdot \nabla\phi$$

$$-\frac{4\pi K T_e n_0}{\omega^2}\nabla \cdot [\boldsymbol{\mu}_e \cdot \nabla(\nabla \cdot \boldsymbol{\mu}_e \cdot \nabla\phi)] = 0 \tag{219}$$

where,

$$\boldsymbol{\mu}_s = \frac{q_s}{m_s}\begin{bmatrix} \dfrac{i\omega}{\Omega_s^2 - \omega^2} & \dfrac{\Omega_s}{\Omega_s^2 - \omega^2} & 0 \\ \dfrac{-\Omega_s}{\Omega_s^2 - \omega^2} & \dfrac{i\omega}{\Omega_s^2 - \omega^2} & 0 \\ 0 & 0 & \dfrac{-i}{\omega} \end{bmatrix}.$$

In order to calculate the ponderomotive force exerted by the lower-hybrid waves, let us assume that the electric field E associated with the latter has $E_x \gg E_z$. Assuming $\mathbf{E} = \tilde{\mathbf{E}}(\mathbf{x}, t)e^{i\omega t}$ where $\tilde{\mathbf{E}}(\mathbf{x}, t)$ is slowly-varying in t, and using (166) for the electrons moving in lower-hybrid electric field, the potential associated with the ponderomotive force is calculated from the averaging over period of $0(\omega^{-1})$ of the nonlinear term $(m_e\mathbf{V}_e \cdot \nabla\mathbf{V}_e)$ to be

$$\psi_e = \frac{e}{4m_e}\frac{|\tilde{E}_x|^2}{\omega^2 - \Omega_e^2}. \tag{220}$$

Assuming that the electrons respond adiabatically to the ponderomotive force exerted by the lower-hybrid waves, one has

$$n_e = n_0 \exp\left[\frac{e}{KT_e}(\varphi - \psi_e)\right] \tag{221}$$

where φ is the low-frequency ambipolar potential. The ions do not move freely along the magnetic field and are coupled to the electrons through the ambipolar potential φ. One has for the ions,

$$n_i = n_0 \exp - \left(\frac{e\varphi}{KT_i}\right). \tag{222}$$

Then, the charge neutrality in the low-frequency motion, namely, $\delta n_e \approx \delta n_i$, gives

$$\varphi = \frac{T_i}{T_i + T_e}\psi_e. \tag{223}$$

Using (223), (221) gives

$$\begin{aligned} \delta n_e &= -n_0 \frac{e\psi_e}{KT} \\ &= -\frac{\omega_{p_e}^2}{16\pi KT}\frac{|\tilde{E}_x|^2}{\omega^2 - \Omega_e^2} \end{aligned} \tag{224}$$

where $T = T_i + T_e$. Using (224), equation (219) becomes

$$\begin{aligned} &K_\perp \frac{\partial^2\phi}{\partial x^2} - K_\parallel \frac{\partial^2\phi}{\partial z^2} + a\frac{\partial^4\phi}{\partial x^4} + b\frac{\partial^4\phi}{\partial x^2\partial z^2} + c\frac{\partial^4\phi}{\partial z^4} \\ &\quad - \alpha\frac{\partial}{\partial x}\left[\frac{\omega_{p_e}^2}{\omega^2 - \Omega_e^2}\frac{|\partial\phi/\partial x|^2}{16\pi n_0 KT}\frac{\partial\phi}{\partial x}\right] \\ &\quad + \beta\frac{\partial}{\partial z}\left[\frac{\omega_{p_e}^2}{\omega^2 - \Omega_e^2}\frac{|\partial\phi/\partial x|^2}{16\pi n_0 KT}\frac{\partial\phi}{\partial z}\right] = 0 \end{aligned} \tag{225}$$

where,

$$K_\perp \equiv 1 + \frac{\omega_{p_e}^2}{\Omega_e^2 - \omega^2} + \frac{\omega_{p_i}^2}{\Omega_i^2 - \omega^2}, \quad K_\parallel = \frac{\omega_{p_e}^2 + \omega_{p_i}^2}{\omega^2} - 1$$

$$a \equiv \frac{\omega_{p_e}^2 V_{T_e}^2}{(\Omega_e^2 - \omega^2)^2}, \quad b \equiv -\frac{2\omega_{p_e}^2 V_{T_e}^2}{\omega^2(\Omega_e^2 - \omega^2)}, \quad c \equiv \frac{\omega_{p_e}^2 V_{T_e}^2}{\omega^4}$$

$$\alpha \equiv \frac{\omega_{p_e}^2}{\Omega_e^2 - \omega^2} + \frac{\omega_{p_i}^2}{\Omega_i^2 - \omega^2}, \quad \beta \equiv \frac{\omega_{p_e}^2}{\omega^2} + \frac{\omega_{p_i}^2}{\omega^2}.$$

The fourth term on the lefthand side in equation (225) (which is of the same order of magnitude as the other terms) was missed by Morales and Lee [89].

Note from equation (225) that in the absence of the nonlinearities and thermal dispersion, the lower-hybrid waves propagate in the direction given by $dx/dz = (K_\perp/K_\parallel)^{1/2}$, which defines the lower-hybrid resonance cone. The latter restricts the region of energy transport; (the group velocity of the lower-hybrid waves is along the cone). The nonlinear effects distort this cone.

Introducing a small parameter $\varepsilon \ll 1$, and

$$\begin{aligned} &\xi = \sqrt{\varepsilon}(x - \lambda z), \quad \eta = \varepsilon^{3/2}\lambda z \\ &u \equiv \sqrt{\varepsilon}\,\frac{\partial \phi}{\partial \xi} \end{aligned} \tag{226}$$

equation (215) becomes

$$\begin{aligned} &2\,\frac{\partial u}{\partial \eta} + \frac{3\omega_{p_e}^2}{\omega^2 - \Omega_e^2}\,\frac{\lambda^2\beta - \alpha}{16\pi n_0 KTK_\perp}\,u^2\,\frac{\partial u}{\partial \xi} \\ &\quad + \frac{1}{K_\perp}(a + b\lambda^2 + c\lambda^4)\,\frac{\partial^3 u}{\partial \xi^3} = 0 \end{aligned} \tag{227}$$

where we have used the result that for the linear case with a cold plasma (which prevails for the $0(\varepsilon)$ problem), $\lambda^2 = K_\perp/K_\parallel$. Equation (227) describes the upper resonant cone; one may obtain a description for the lower resonant cone in a similar manner by introducing instead

$$\xi = \sqrt{\varepsilon}\,(x + \lambda z), \quad \eta = -\varepsilon^{3/2}\lambda z. \tag{228}$$

Equation (227) is a modified Korteweg–de Vries equation (Hirota [187]), and has solitons as localized solutions. The latter have intense localized electric fields associated with them.

It may be mentioned that there are some issues regarding the existence of solutions of equation (227) for physically realizable boundary conditions (Karney *et al.* [188, 189]).

(v) Whistler Solitons

Self-modulation of whistler waves has been known to lead to localization of these waves. The localized waves consist of quasi-stationary envelopes of whistler electromagnetic fields accompanied by either density dips or humps, depending on the frequency of the whistler waves (Karpman and Washimi [94, 95]). The self-focusing processing is believed to be the result of the reinforcing interaction between the wave-induced density depression and the density-induced wave refraction. The evolution of the slowly-varying envelope of a high-frequency whistler wave is described by the nonlinear Schrödinger equation. The ponderomotive force executed by the whistler field induces a plasma motion which is sufficiently slow and weak so that it can be described by linearized ideal MHD equations. Here we restrict ourselves to the case where the whistler wave propagates along the applied magnetic field ($\mathbf{B}_0 = B_0\hat{\mathbf{i}}_z$) and all variables depend only on z and t. The following version of analysis is due to Mann and Motschmann [190].

Let us first obtain an equation governing the envelope of finite-amplitude whistler waves in the presence of slow modulations of the background density and magnetic field. Let the electric field pulse be given by $\frac{1}{2}\mathbf{E}(z, t)$ $\exp[i(\mathbf{k}\cdot\mathbf{x} - \omega t) + \text{c.c.}$, where $\mathbf{E}(z, t)$ is a slowly-varying function of z and t. Then, in the complex representation $E = E_x + iE_y$ of a right-circularly polarized wave, the equation for the evolution of the whistler electric field envelope is

$$i\left(\frac{\partial E}{\partial t} + V_g\frac{\partial E}{\partial z}\right) + \frac{1}{2}V_g'\frac{\partial^2 E}{\partial z^2} + (\omega_0 - \omega)E = 0 \tag{229}$$

where,

$$\omega = \frac{c^2k^2\Omega_e(B)}{\omega_{p_e}^2(n) + c^2k^2} + kV_z \tag{230}$$

$$V_g = \frac{\partial\omega_0}{\partial k} = 2(1-\alpha)\frac{\omega_0}{k}, \quad \alpha \equiv \frac{\omega_0}{\Omega_e}$$
$$V_g' = \frac{\partial^2\omega_0}{\partial k^2} = 2(1-\alpha)(1-4\alpha)\frac{\omega_0}{k^2}. \tag{231}$$

The electron plasma frequency $\omega_{p_e}(n)$ and the gyrofrequency $\Omega_e(B_0)$ are to be evaluated at the local values of the density n and the magnetic field $\mathbf{B}_0$, while the constant ω_0 is the whistler frequency ω evaluated at ambient ($x = \pm\infty$) conditions. Let us write

$$n = n_0 + \tilde{n} \tag{232}$$

where n_0 is the unperturbed density, and assume that the background response to the large-amplitude whistler waves is electrostatic and is governed by the linearized MHD equations:

$$\frac{\partial\tilde{n}}{\partial t} + n_0\frac{\partial V_z}{\partial z} = 0 \tag{233}$$

$$\frac{\partial V_z}{\partial t} = -\frac{C_s^2}{n_0}\frac{\partial\tilde{n}}{\partial z} + \frac{1}{m_i n_0}f_z \tag{234}$$

where $C_s^2 \equiv KT_e/m_i$, T_e being the electron temperature and f_z is the ponderomotive force excited by the high-frequency whistler field (see the Appendix to this chapter):

$$f_z = \frac{\omega_{p_e}^2}{\omega_0(\Omega_e - \omega_0)}\left(\frac{\partial}{\partial z} + \frac{2}{V_g}\frac{\partial}{\partial t}\right)\frac{|\mathbf{E}|^2}{16\pi}. \tag{235}$$

Using (230) and (232), equation (229) becomes

$$i\left(\frac{\partial E}{\partial t}+V_g\frac{\partial E}{\partial z}\right)+\frac{1}{2}V_g'\frac{\partial^2 E}{\partial z^2}$$

$$+\Omega_e\alpha(1-\alpha)\left(\frac{\tilde{n}}{n}-\frac{2V_z}{V_g}\right)E=0. \tag{236}$$

Assuming that the slow response evolves according to a stationary propagation with velocity V, equations (233)–(235) give

$$n_0V_z = V\tilde{n} \tag{237}$$

$$\tilde{n}=\frac{\omega_{p_e}^2}{\omega_0(\Omega_e-\omega_0)}\frac{(1-2V/V_g)}{KT_e(1-M^2)}\frac{|\mathbf{E}|^2}{16\pi} \tag{238}$$

where $M \equiv V/C_s$ and the constants of integration are chosen such that n and V_z vanish when $E = 0$.

Using (237) and (238), equation (236) becomes:

$$i\left(\frac{\partial E}{\partial t}+V_g\frac{\partial E}{\partial z}\right)+\frac{1}{2}V_g'\frac{\partial^2 E}{\partial z^2}+$$

$$+\frac{\omega_{p_e}^2/\Omega_e}{n_0KT_e(1-M^2)}\frac{(1-2V/V_g)}{32\pi}|E|^2E=0. \tag{239}$$

Equation (239) has the solution:

$$E(z,t)=A\,\mathrm{sech}\left[\sqrt{\frac{\omega_{p_e}^2k^2(1-2V/V_g)}{n_0KT_e\omega_0\Omega_e(1-M^2)(1-\alpha)(1-4\alpha)}}\times\right.$$

$$\times A(z-V_gt)\times\exp\left\{i\left[\frac{V-V_g}{V_g'}z-\right.\right.$$

$$\left.\left.-\left\{\frac{V^2-V_g^2}{2V_g'}-\frac{A^2}{64\pi}\frac{\omega_{p_e}^2/\Omega_e^2(1-2V/V_g)}{n_0KT_e(1-M^2)}\right\}t\right]\right\} \tag{240}$$

which represents an envelope soliton. Observe from (240) that one has supersonic solitons for $\alpha < \frac{1}{4}$ and subsonic solitons for $\alpha > \frac{1}{4}$ (note that one requires $V \approx V_g$ in order that equation (239) remains valid). Equation (238) then shows that one has density humps for supersonic solitons and density dips for subsonic solitons.

Stenzel [191] made observations of a filamentation instability of a large amplitude whistler wave in a collisionless, quiescent plasma column. The wave was launched in a large uniform magnetoplasma from antennas which produced a diverging energy flow in the linear regime. At large amplitudes, the radiation pressure gave rise to a density depression originating from the antenna. The wave refracted into the density trough which locally enhanced the wave intensity and a filamentation instability developed. Figure 5.17 shows a comparison of the wave intensity vs. radial position at different axial distances from the exciter for a linear (Figure 5.17a) and nonlinear (Figure 5.17b) whistler

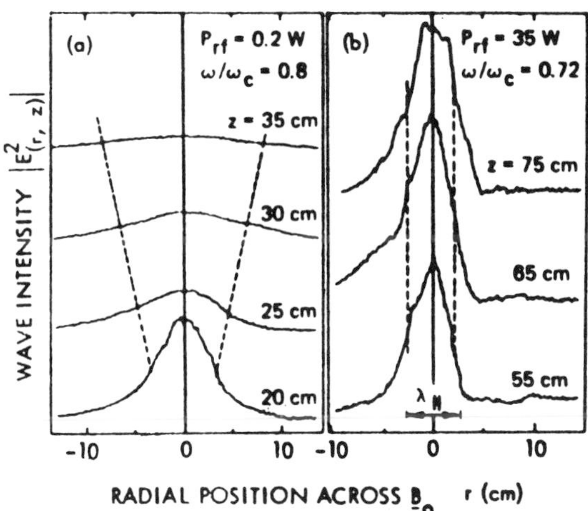

Figure 5.17. Wave intensity across the magnetic field at different axial distances from the dipole exciter antenna. In the linear regime (a) the wave intensity decreases rapidly with distance z due to the diverging energy flow. In the nonlinear regime (b) the wave intensity is confined to a narrow, field-aligned nondiverging filament. (Due to Stenzel [191], by courtesy of The American Institute of Physics.)

wave. While the small-amplitude wave has a diverging energy flow, the large-amplitude wave is confined to a narrow region across $\mathbf{B}_0$ which does not broaden with increasing distance from the exciter antenna. Thus, the large-amplitude wave has created a duct or filament in which the wave energy is confined. The duct formation, on the other hand, implies that the background plasma parameters have been changed by the wave. Time- and space-resolved density measurements indeed showed that the duct represented a field-aligned density depression. Figure 5.18a shows the radial density profile sampled at different times t_s after the large-amplitude whistler wave has been turned on. As time goes on, an initially flat profile is seen to evolve into a density depression which deepens and widens. The density minimum corresponded to the location of the antenna, the half-width was wider but of the order of the antenna dimensions across $\mathbf{B}_0$. The axial profile (Figure 5.18b) shows that the density depression originates from the antenna and subsequently spreads out along $\mathbf{B}_0$ to form eventually a long field aligned density trough. The filamentation instability was convective in the direction of $\mathbf{B}_0$. As a consequence, the instability stopped when the duct reached the axial plasma boundaries.

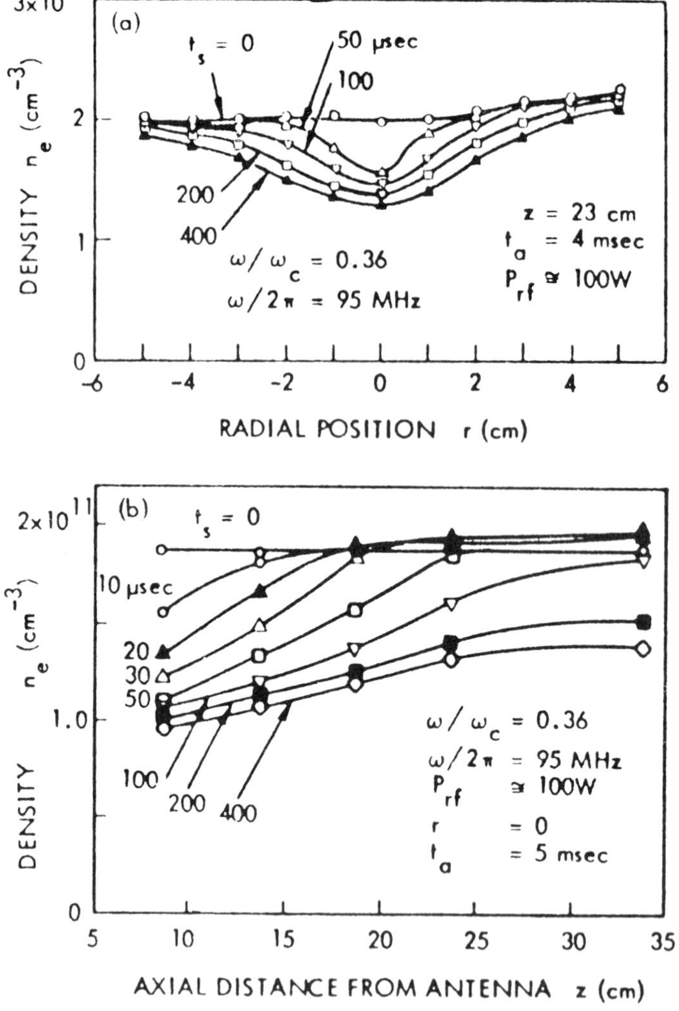

Figure 5.18. Density profiles across (a) and along (b) the magnetic field sampled at different times t_s after the turn-on of the large amplitude wave applied to a 1.6 cm diam. magnetic loop antenna at $r = 0$, $z = 0$. (Due to Stenzel [191], by courtesy of The American Institute of Physics.)

Appendix

We will calculate the time derivative term in the ponderomotive force which arises from averaging over the last oscillations of the term $\mathbf{F} = c^{-1}(\mathbf{J} \times \mathbf{H})$. We have

$$\mathbf{J} = \frac{1}{4\pi} \frac{\partial}{\partial t} (\mathbf{D} - \mathbf{E})$$

$$= \frac{1}{4\pi} \frac{\partial}{\partial t} \int [\boldsymbol{\varepsilon}(\omega + \mu) - \mathbf{I}] \cdot \mathbf{E}(\mu) \exp[-i(\omega + \mu)t] \, d\mu \tag{A.1}$$

where μ is the Fourier transform of the slow-time variable and we have written $\mathbf{E} = \mathbf{E}(t)e^{-i\omega t}$. On expanding $\boldsymbol{\varepsilon}(\omega + \mu)$, we obtain to $0(\mu)$,

$$\mathbf{J} = -\frac{i}{4\pi}\int(\omega+\mu)\left[\boldsymbol{\varepsilon}(\omega)+\frac{\partial\boldsymbol{\varepsilon}}{\partial\omega}\mu-\mathbf{I}\right]\cdot\mathbf{E}(\mu)\times\exp[-i(\omega+\mu)t]\,\mathrm{d}\mu$$

$$\approx\frac{-i}{4\pi}\left[\omega\left\{\boldsymbol{\varepsilon}(\omega)-\mathbf{I}\right\}\cdot\mathbf{E}(t)e^{-i\omega t}\right.$$

$$\left.+\int\mu\left(\omega\frac{\partial\boldsymbol{\varepsilon}}{\partial\omega}+\boldsymbol{\varepsilon}-\mathbf{I}\right)\cdot\mathbf{E}(\mu)\exp[-i(\omega+\mu)t]\,\mathrm{d}\mu\right]$$

$$=\frac{1}{4\pi}\left[-i\omega\left\{\boldsymbol{\varepsilon}(\omega)-\mathbf{I}\right\}+\left\{\frac{\partial(\omega\boldsymbol{\varepsilon})}{\partial\omega}-\mathbf{I}\right\}\frac{\partial}{\partial t}\right]\cdot\mathbf{E}(t)e^{-i\omega t}. \tag{A.2}$$

The above derivation was given by Spatschek *et al.* [191].

Thus, the time-dependent part of **F** is given by

$$\mathbf{F}_t=\frac{1}{c}\langle\mathbf{J}\times\mathbf{H}\rangle$$

$$=\frac{1}{16\pi c}\left[\left\{\frac{\partial(\omega\boldsymbol{\varepsilon})}{\partial\omega}-\mathbf{I}\right\}\cdot\frac{\partial\mathbf{E}}{\partial t}\times\mathbf{H}^*-i\omega(\boldsymbol{\varepsilon}-\mathbf{I})\cdot\mathbf{E}\times\mathbf{H}^*+\text{c.c.}\right]. \tag{A.3}$$

Using the relation,

$$\mathbf{H}=\frac{i}{\omega}\left(c\nabla\times\mathbf{E}+\frac{\partial\mathbf{H}}{\partial t}\right) \tag{A.4}$$

where the second term is due to the slowly-varying part of **H**, the time-dependent part (A.3) becomes

$$\mathbf{F}_t=\frac{1}{16\pi c}\left[\left\{\frac{\partial(\omega\boldsymbol{\varepsilon})}{\partial\omega}-\mathbf{I}\right\}\cdot\frac{\partial\mathbf{E}}{\partial t}\times\mathbf{H}^*+(\boldsymbol{\varepsilon}-\mathbf{I})\cdot\mathbf{E}\times\frac{\partial\mathbf{H}^*}{\partial t}+\text{c.c.}\right]$$

$$=\frac{1}{16\pi c}\left[\frac{\partial}{\partial t}\left\{(\boldsymbol{\varepsilon}-\mathbf{I})\cdot\mathbf{E}\times\mathbf{H}^*\right\}+\omega\frac{\partial\boldsymbol{\varepsilon}}{\partial\omega}\cdot\frac{\partial\mathbf{E}}{\partial t}\times\mathbf{H}^*+\text{c.c.}\right]. \tag{A.5}$$

If the whistler field is a quasi-plane wave with wave vector **k**, then we have (on assuming $|\boldsymbol{\varepsilon}|\gg 1$),

$$F_{t_z}=\frac{k}{16\pi\omega^2}\frac{\partial(\omega^2\varepsilon)}{\partial\omega}\frac{\partial}{\partial t}|\mathbf{E}|^2 \tag{A.6}$$

where,

$$\varepsilon_{ij}=\varepsilon\delta_{ij},\quad \varepsilon=\frac{\omega_{p_e}^2}{\omega(\Omega_e-\omega)}\gg 1.$$

Thus, we have for the total ponderomotive force

$$F_z = \left[\varepsilon \frac{\partial}{\partial z} + \frac{k}{\omega^2} \frac{\partial(\omega^2 \varepsilon)}{\partial \omega} \frac{\partial}{\partial t} \right] \frac{|\mathbf{E}|^2}{16\pi} . \tag{A.7}$$

Noting that,

$$\frac{k}{\omega^2} \frac{\partial(\omega^2 \varepsilon)}{\partial \omega} = \frac{2\varepsilon}{V_g}$$

we obtain (235).

VI

Nonlinear Relativistic Waves

When an electromagnetic wave propagating through a plasma is strong enough to drive the electrons to relativistic speeds, two nonlinear effects come into play (Akhiezer and Polovin [101], Lunow [102], Kaw and Dawson [103], Max and Perkins [104], Chian and Clemmow [105], and DeCoster [106], Kaw *et al.* [107])

(i) relativistic variation of the electron mass;
(ii) excitation of space-charge fields by strong $\mathbf{V} \times \mathbf{B}$ forces driving electrons along the direction of propagation of the electromagnetic wave.

The first effect leads to a propagation of the electromagnetic wave in a normally overdense plasma, and the second effect leads to a coupling of the electromagnetic wave to the Langmuir wave in the plasma.

(i) A Coupled System of an Electromagnetic Wave and a Langmuir Wave

Consider an intense electromagnetic wave propagating along the z-axis in a plasma. Under the action of this wave field, let us assume that the electrons form a cold fluid (i.e., their directed speeds are much larger than their random speeds) undergoing nonlinear oscillations, whereas, the ions make an immobile neutralizing positive-charge background. The basic equations describing the propagation of this electromagnetic wave coupled nonlinearly to the Langmuir wave in the plasma are

$$\nabla \times \mathbf{E} = -\frac{1}{c}\frac{\partial \mathbf{B}}{\partial t} \tag{1}$$

$$\nabla \times \mathbf{B} = \frac{1}{c}\frac{\partial \mathbf{E}}{\partial t} - \frac{4\pi}{c} ne\mathbf{V} \tag{2}$$

$$\nabla \cdot \mathbf{E} = -4\pi e(n - n_0) \tag{3}$$

$$\nabla \cdot \mathbf{B} = 0 \tag{4}$$

$$\frac{\partial \mathbf{p}}{\partial t} + (\mathbf{V} \cdot \nabla)\mathbf{p} = -e\left[\mathbf{E} + \frac{1}{c}\mathbf{V} \times \mathbf{B}\right] \tag{5}$$

where,

$$\mathbf{p} = \frac{m\mathbf{V}}{\sqrt{1 - V^2/c^2}} \tag{6}$$

m being the rest-mass of an electron.

Let us choose the electromagnetic field for the coupled system with a polarization shown in Figure 6.1. Assuming that all quantities depend only on z and t, we have the equations (1)—(5):

$$\frac{\partial E_x}{\partial z} = -\frac{1}{c}\frac{\partial B_y}{\partial t} \tag{7}$$

$$\frac{\partial E_y}{\partial z} = \frac{1}{c}\frac{\partial B_x}{\partial t} \tag{8}$$

$$\frac{\partial B_x}{\partial z} = \frac{1}{c}\frac{\partial E_y}{\partial t} - \frac{4\pi}{c} neV_y \tag{9}$$

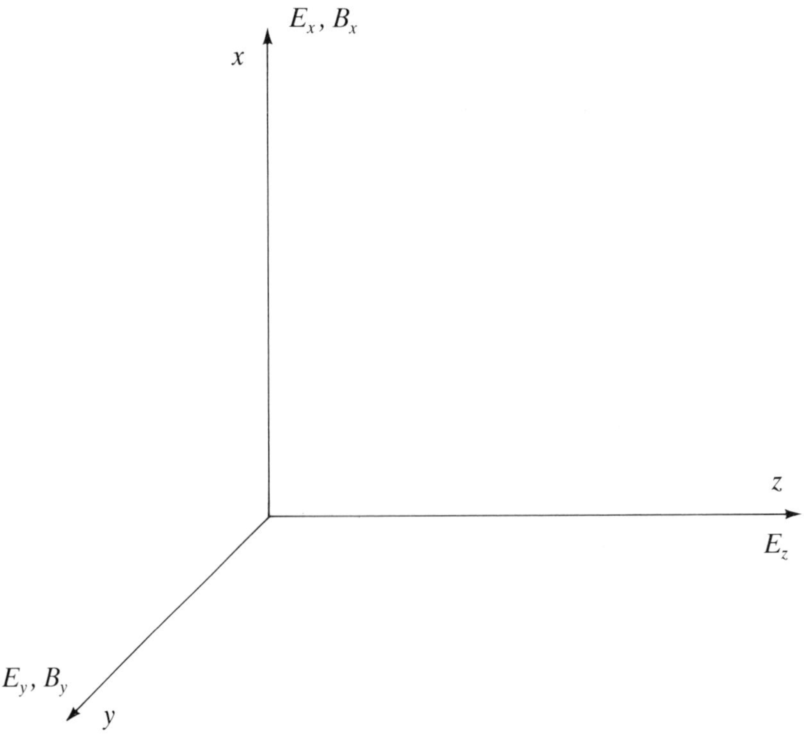

Figure 6.1. Electromagnetic field polarization.

$$-\frac{\partial B_y}{\partial z} = \frac{1}{c}\frac{\partial E_x}{\partial t} - \frac{4\pi}{c}\, neV_x \tag{10}$$

$$\frac{1}{c}\frac{\partial E_z}{\partial t} = \frac{4\pi}{c}\, neV_z \tag{11}$$

$$\frac{\partial E_z}{\partial z} = -4\pi e(n - n_0) \tag{12}$$

$$\left(\frac{\partial}{\partial t} + V_z\frac{\partial}{\partial z}\right)\left(\frac{mV_x}{\sqrt{1 - V^2/c^2}}\right) = -eE_x + \frac{e}{c}\, V_zB_y \tag{13}$$

$$\left(\frac{\partial}{\partial t} + V_z\frac{\partial}{\partial z}\right)\left(\frac{mV_y}{\sqrt{1 - V^2/c^2}}\right) = -eE_y + \frac{e}{c}\, V_zB_x \tag{14}$$

$$\left(\frac{\partial}{\partial t} + V_z\frac{\partial}{\partial z}\right)\left(\frac{mV_z}{\sqrt{1 - V^2/c^2}}\right) = -eE_z + \frac{e}{c}\,(V_yB_x - V_xB_y). \tag{15}$$

Let us seek plane nonlinear travelling wave-solutions of equations (7)—(15), of the form

$$q(z, t) = q(\eta), \quad \eta \equiv z - ut \tag{16}$$

then equations (7)—(12) give:

$$\begin{gathered} E_x = \frac{u}{c}\, B_y, \quad E_y = -\frac{u}{c}\, B_x, \quad E_z' = -\frac{4\pi n_0 e}{u - V_z}\, V_z \\ n = \frac{n_0 u}{u - V_z}, \quad B_x' = \frac{4\pi n_0 e}{c(\beta^2 - 1)}\, \frac{uV_y}{u - V_z}, \\ B_y' = -\frac{4\pi n_0 e}{c(\beta^2 - 1)}\, \frac{uV_x}{u - V_z} \end{gathered} \tag{17}$$

where primes denote differentiation with respect to η, and $\beta \equiv u/c$.

Using (17), and introducing

$$\mathbf{P} = \frac{\mathbf{V}/c}{\sqrt{1 - V^2/c^2}} \tag{18}$$

one may derive from equations (13)—(15):

$$P_x'' + \frac{\omega_p^2}{c^2}\, \frac{\beta P_x}{(\beta^2 - 1)\,[\beta\sqrt{1 + P^2} - P_z]} = 0 \tag{19}$$

$$P_y'' + \frac{\omega_p^2}{c^2}\, \frac{\beta P_y}{(\beta^2 - 1)\,[\beta\sqrt{1 + P^2} - P_z]} = 0 \tag{20}$$

$$(\beta P_z - \sqrt{1 + P^2})'' + \frac{\omega_p^2}{c^2} \frac{P_z}{[\beta\sqrt{1 + P^2} - P_z]} = 0. \tag{21}$$

Analytical solutions of these equations have not been obtained for the general case of coupled electromagnetic-Langmuir waves. Exact solutions have been obtained only for some special case like purely transverse waves and purely longitudinal waves. For the general case, we can give only some approximate solutions using a perturbation theory.

(a) Purely Transverse Waves

Setting $P_z = 0$, one obtains from equation (21), P^2 = constant, and using equations (19) and (20), one obtains

$$\begin{aligned} P_x &= P \cos \omega\eta \\ P_y &= P \sin \omega\eta \end{aligned} \tag{22}$$

where,

$$\omega = \frac{\beta\omega_p}{\sqrt{\beta^2 - 1}} \frac{1}{(1 + P^2)^{1/4}}. \tag{23}$$

Equation (22) shows that if the waves are purely transverse, they can only be circularly-polarized. This is because of the fact that for small amplitudes, one can superpose two oscillations of opposite circular polarization leading to the appearance of linearly polarized oscillation, but this superposition is not valid for waves of large amplitudes.

Using (23), one may write

$$u = \beta c = \frac{c}{\sqrt{\varepsilon}} \tag{24}$$

where,

$$\varepsilon = 1 - \frac{\omega_p^2/\omega^2}{\sqrt{1 + P^2}}. \tag{25}$$

Now, noting that

$$E_x \approx \frac{mc\omega P}{e} \sin \omega\eta, \quad E_y \approx \frac{mc\omega P}{e} \cos \omega\eta$$

or

$$P^2 = \frac{e^2 E_0^2}{m^2\omega^2c^2}, \quad E_0^2 = E_x^2 + E_y^2.$$

Equation (25) can be written as

$$\varepsilon = 1 - \frac{\omega_p^2/\omega^2}{(1 + e^2E_0^2/m^2c^2\omega^2)^{1/2}}. \tag{26}$$

Noting that one requires for propagation $\varepsilon > 0$, one sees from (26) that waves in the frequency range $[\omega_p/(1 + e^2E_0^2/m^2c^2\omega^2)^{1/2}] < \omega < \omega_p$ can also propagate in the plasma. This is in contrast to what would be expected from linear theory according to which waves with $\omega < \omega_p$ cannot propagate. Propagation in overdense plasmas may be understood by noting that the electrons become more massive and the effective plasma frequency is reduced so that the waves propagate as long as their frequency is above this effective plasma frequency. Alternately, one may note that the plasma current is limited (to nec) instead of increasing with E as $nev = ne^2E/m\omega$ so that it does not become large enough to cancel the displacement current and hence there is no wave reflection. Relativistic effects thus diminish the ability of the plasma to act as a dielectric.

(b) Luminous Waves

In order to treat waves with $\beta \cong 1$, let us introduce

$$\theta = \frac{\omega_p}{c} \frac{\eta}{\sqrt{\beta^2 - 1}}. \tag{27}$$

Equations (19)–(21) then become

$$\frac{d^2P_x}{d\theta^2} + \frac{P_x}{\sqrt{1 + P^2} - P_z} = 0 \tag{28}$$

$$\frac{d^2P_y}{d\theta^2} + \frac{P_y}{\sqrt{1 + P^2} - P_z} = 0 \tag{29}$$

$$\frac{d^2}{d\theta^2}[P_z - \sqrt{1 + P^2}] + (\beta^2 - 1)P_z[\sqrt{1 + P^2} - P_z] = 0. \tag{30}$$

Equation (30) gives

$$\sqrt{1 + P^2} - P_z \approx \text{const} = C^2. \tag{31}$$

Using (31), equations (28) and (29) give

$$\begin{aligned} P_x &= R_x \cos \frac{\theta}{C} \\ P_y &= R_y \sin \frac{\theta}{C}. \end{aligned} \tag{32}$$

Using (32), (31) gives

$$P_z = \frac{1}{4C^2}\left[R_x^2 + R_y^2 - 2(C^4 - 1) + (R_x^2 - R_y^2)\cos\frac{2\theta}{C}\right]. \tag{33}$$

Averaging over one wave period, equation (11) gives for the average

$$\langle nV_z \rangle = 0 \tag{34}$$

so that from (34)

$$\langle P_z \rangle = 0. \tag{35}$$

Using (33), (34) then gives

$$R_x^2 + R_y^2 = 2(C^4 - 1). \tag{36}$$

Using (36), (32) and (33) become

$$\begin{aligned} P_x &= R_x \cos \omega\eta, \quad P_y = R_y \sin \omega\eta \\ P_z &= \frac{\frac{1}{4}(R_x^2 - R_y^2)}{[1 + \frac{1}{2}(R_x^2 + R_y^2)]^{1/2}} \cos 2\omega\eta \end{aligned} \tag{37}$$

where

$$\omega = \frac{\omega_p/c}{(\beta^2 - 1)^{1/2}[1 + \frac{1}{2}(R_x^2 + R_y^2)]^{1/4}}.$$

(c) A Perturbation Theory for the General Coupled System

Most of the work on this problem is concerned with obtaining exact (or nearly exact) analytical solutions for some special cases like purely transverse waves and purely longitudinal waves, as we discussed in the above. DeCoster [106] considered the general case, for which he gave some approximate solutions. However, DeCoster's procedure does not allow for amplitude modulations of the coupled waves, and is therefore restricted in scope. Actually, DeCoster's calculation will be invalid, if the system of coupled waves in question exhibits internal resonances with the concomitant modulations in the amplitudes of the coupled waves. We will now give below a generalized perturbation theory (Shivamoggi [63]) that allows for both amplitude and phase modulations of the coupled waves. This theory can therefore successfully deal with internal resonances if they arise in the system in question. This theory also recovers results of the known special cases in the appropriate limit.

For the general case, let us consider the electromagnetic wave to be linearly polarized so that $P_y = 0$. On introducing

$$\begin{aligned} \chi &= \sqrt{\beta^2 - 1}\,P_x, \quad Z = \beta P_z - \sqrt{1 + P^2} \\ \xi &= \frac{\omega_p}{c} \frac{\eta}{\sqrt{\beta^2 - 1}} \end{aligned} \tag{38}$$

one obtains from equations (18) and (20):

$$\frac{d^2\chi}{d\xi^2} + \frac{\beta\chi}{[\beta^2 - 1 + \chi^2 + Z^2]^{1/2}} = 0 \tag{39}$$

$$\frac{d^2Z}{d\xi^2} + \frac{\beta Z}{[\beta^2 - 1 + \chi^2 + Z^2]^{1/2}} = 0. \tag{40}$$

Let us now assume that the coupling between the electromagnetic wave and the Langmuir wave is weak, and let a small parameter ε characterize the magnitude of this weak coupling. Let us put,

$$\chi = \varepsilon X, \quad Z = -1 + \varepsilon Z. \tag{41}$$

Equations (39) and (40) then give on expansion of the radicals:

$$\frac{d^2X}{d\xi^2} + X = -\varepsilon \frac{1}{\beta^2} XZ + \varepsilon^2 \frac{1}{2\beta^2}\left[X^3 + \left(1 - \frac{3}{\beta^2}\right)XZ^2\right] + \cdots \tag{42}$$

$$\begin{aligned}\frac{d^2Z}{d\xi^2} + \left(1 - \frac{1}{\beta^2}\right)Z &= -\varepsilon\left[\frac{3}{2\beta^2}\left(1 - \frac{1}{\beta^2}\right)Z^2 + \frac{X^2}{2\beta^2}\right]\\ &+ \varepsilon^2 \frac{1}{2\beta^2}\left[\left(1 - \frac{6}{\beta^2} + \frac{5}{\beta^4}\right)Z^3 + \left(1 - \frac{3}{\beta^2}\right)X^2Z\right] + \cdots.\end{aligned} \tag{43}$$

Note that X corresponds to the electromagnetic wave, and Z corresponds to the Langmuir wave.

Let us seek solutions to equations (42) and (43) of the form:

$$\begin{aligned}X &= A_1(\xi_1, \xi_2)\cos\phi_1(\xi_1, \xi_2) + \varepsilon u_1 + \varepsilon^2 u_2 + \cdots\\ Z &= A_2(\xi_1, \xi_2)\cos\phi_2(\xi_1, \xi_2) + \varepsilon v_1 + \varepsilon^2 v_2 + \cdots\end{aligned} \tag{44}$$

where,

$$\begin{aligned}&\phi_1(\xi_1, \xi_2) = \xi - \theta_1(\xi_1, \xi_2)\\ &\phi_2(\xi_1, \xi_2) = \sqrt{1 - 1/\beta^2}\,\xi - \theta_2(\xi_1, \xi_2)\\ &\xi_1 \equiv \varepsilon\xi, \quad \xi_2 \equiv \varepsilon^2\xi.\end{aligned} \tag{45}$$

The motivation behind the prescription (44) is the expectation that solutions of equations (42) and (43) for $\varepsilon \neq 0$ (but for small ε) are very nearly simple harmonics of the form $A_1 \cos\phi_1$, $A_2 \cos\phi_2$ which they would identically be if $\varepsilon = 0$. The perturbations induced by the terms with $\varepsilon \neq 0$ on the right hand side of equations (42) and (43) are then expected to be reflected in slow changes (characterized by the slow scales ξ_1, ξ_2) in the amplitudes A_1, A_2 and phases θ_1, θ_2 of the near harmonics and in higher harmonics through the u's and v's.

Using (44) and (45), we obtain from equations (42) and (43):

$$(2\varepsilon A_1\theta_{1_{\xi_1}} + 2\varepsilon^2 A_1\theta_{1_{\xi_2}} - \varepsilon^2 A_1\theta^2_{1_{\xi_1}} + \varepsilon^2 A_{1_{\xi_1\xi_1}})\cos\phi_1$$

$$+ (-2\varepsilon A_{1_{\xi_1}} - 2\varepsilon^2 A_{1_{\xi_2}} + 2\varepsilon^2 A_{1_{\xi_1}}\theta_{1_{\xi_1}} + \varepsilon^2 A_1\theta_{1_{\xi_1\xi_1}})\sin\phi_1$$

$$+ \varepsilon(u_{1_{\xi\xi}} + u_1) + \varepsilon^2(2u_{1_{\xi\xi_1}} + u_{2_{\xi\xi}} + u_2) + \cdots$$

$$= -\varepsilon\frac{1}{\beta^2}(A_1 A_2\cos\phi_1\cdot\cos\phi_2)$$

$$+ \varepsilon^2\frac{1}{\beta^2}\left[-(A_1\cos\phi_1\cdot v_1 + A_2\cos\phi_2\cdot u_1)\right.$$

$$\left. + \frac{1}{2}\left\{A_1^3\cos^3\phi_1 + \left(1 - \frac{3}{\beta^2}\right)A_1 A_2^2\cos\phi_1\cdot\cos^2\phi_2\right\}\right] + \cdots \tag{46}$$

$$(2\varepsilon\sqrt{1 - 1/\beta^2}\,A_2\theta_{2_{\xi_1}} + 2\varepsilon^2\sqrt{1 - 1/\beta^2}\,A_2\theta_{2_{\xi_2}} - \varepsilon^2 A_2\theta^2_{2_{\xi_1}}$$

$$+ \varepsilon^2 A_{2_{\xi_1\xi_1}})\cos\phi_2 + (-2\varepsilon\sqrt{1 - 1/\beta^2}\,A_{2_{\xi_1}}$$

$$- 2\varepsilon^2\sqrt{1 - 1/\beta^2}\,A_{2_{\xi_2}} + 2\varepsilon^2 A_{2_{\xi_1}}\theta_{2_{\xi_1}} + \varepsilon^2 A_2\theta_{2_{\xi_1\xi_1}})\sin\phi_2$$

$$+ \varepsilon[v_{1_{\xi\xi}} + (1 - 1/\beta^2)v_1] + \varepsilon^2[2v_{1_{\xi\xi_1}} + v_{2_{\xi\xi}} + (1 - 1/\beta^2)v_2] + \cdots$$

$$= -\varepsilon\frac{1}{2\beta^2}[3(1 - 1/\beta^2)A_2^2\cos^2\phi_2 + A_1^2\cos^2\phi_1]$$

$$+ \varepsilon^2\frac{1}{\beta^2}\left[-\left\{\frac{3}{2}(1 - 1/\beta^2)A_2\cos\phi_2\cdot v_1 + A_1\cos\phi_1\cdot u_1\right\}\right.$$

$$+ \frac{1}{2}\left\{\left(1 - \frac{6}{\beta^2} + \frac{5}{\beta^4}\right)A_2^3\cos^3\phi_2\right.$$

$$\left.\left. + (1 - 3/\beta^2)A_1^2 A_2\cos^2\phi_1\cos\phi_2\right\}\right] + \cdots. \tag{47}$$

By equating the coefficients of $\sin\phi_1$, $\cos\phi_1$, $\sin\phi_2$, $\cos\phi_2$ and the rest to zero separately, one obtains from equations (46) and (47), to $0(\varepsilon)$:

$$\theta_{1_{\xi_1}} = 0 \tag{48}$$

$$A_{1_{\xi_1}} = 0 \tag{49}$$

$$u_{1_{\xi\xi}} + u_1 = -\frac{1}{\beta^2} A_1 A_2 \cos\phi_1 \cdot \cos\phi_2 \tag{50}$$

$$\theta_{2_{\xi_1}} = 0 \tag{51}$$

$$A_{2_{\xi_1}} = 0 \tag{52}$$

$$v_{1_{\xi\xi}} + \left(1 - \frac{1}{\beta^2}\right) v_1 = -\frac{3}{2\beta^2}\left(1 - \frac{1}{\beta^2}\right) A_2^2 \cos^2\phi_2$$
$$- \frac{1}{2\beta^2} A_1^2 \cos^2\phi_1 . \tag{53}$$

On solving equations (48)—(53), one obtains

$$u_1 = \frac{A_1A_2/2\beta^2}{(1 + \sqrt{1 - 1/\beta^2})^2 - 1} \cos(\phi_1 + \phi_2)$$
$$+ \frac{A_1A_2/2\beta^2}{(1 - \sqrt{1 - 1/\beta^2})^2 - 1} \cos(\phi_1 - \phi_2)$$

$$v_1 = \frac{-1/4\beta^2}{(1 - 1/\beta^2)} [3(1 - 1/\beta^2)A_2^2 + A_1^2] + \frac{A_2^2}{4\beta^2} \cos 2\phi_2$$
$$- \frac{A_1^2/4\beta^2}{(1 - 1/\beta^2) - 4} \cos 2\phi_1 . \tag{54}$$

Where the A's and θ's are constants to $0(\varepsilon)$. Thus, in general, the two waves propagate without any appreciable energy exchange between them.

Next, using (48)—(54), one obtains from equations (46) and (47) to $0(\varepsilon^2)$:

$$2A_1\theta_{1_{\xi_2}} \cos\phi_1 - 2A_{1_{\xi_2}} \sin\phi_1 + u_{2_{\xi\xi}} + u_2$$

$$= -\frac{1}{\beta^2}\left[\frac{-1/4\beta^2}{(1 - 1/\beta^2)}\{3(1 - 1/\beta^2)A_2^2 + A_1^2\}\right.$$
$$- \frac{A_1^2/8\beta^2}{(1 - 1/\beta^2) - 4} + \frac{A_2^2/4\beta^2}{(1 + \sqrt{1 - 1/\beta^2})^2 - 1}$$
$$\left. + \frac{A_2^2/4\beta^2}{(1 - \sqrt{1 - 1/\beta^2})^2 - 1} - \frac{3A_1^2}{8} - \frac{1}{4}\left(1 - \frac{3}{\beta^2}\right) A_2^2\right] A_1 \cos\phi_1$$

$$+ \left[-\frac{A_1A_2^2}{8\beta^4} - \frac{A_1A_2^2/4\beta^4}{(1 + \sqrt{1 - 1/\beta^2})^2 - 1} + \frac{1}{8\beta^2}\left(1 - \frac{3}{\beta^2}\right) A_1A_2^2\right] \times$$

$$\times \cos(\phi_1 + 2\phi_2) + \left[-\frac{A_1 A_2^2}{8\beta^4} - \frac{A_1 A_2^2/4\beta^4}{(1-\sqrt{1-1/\beta^2})^2 - 1} \right.$$

$$\left. + \frac{1}{8\beta^2}\left(1 - \frac{3}{\beta^2}\right) A_1 A_2^2 \right] \cos(\phi_1 - 2\phi_2)$$

$$+ \left[\frac{A_1^3}{8\beta^2} + \frac{A_1^3/8\beta^4}{(1-1/\beta^2) - 4} \right] \cos 3\phi_1 \tag{55}$$

$$2A_2\sqrt{1-1/\beta^2}\,\theta_{2_{\xi_2}} \cos\phi_2 - 2A_{2_{\xi_2}}\sqrt{1-1/\beta^2}\sin\phi_2 + v_{2_{\xi\xi}} + (1-1/\beta^2)v_2$$

$$= \left[\frac{3}{4\beta^4}\left\{ 3\left(1 - \frac{1}{\beta^2}\right) A_2^2 + A_1^2 \right\} - \frac{3}{8\beta^4}\left(1 - \frac{1}{\beta^2}\right) A_2^2 \right.$$

$$- \frac{A_1^2/4\beta^4}{(1+\sqrt{1-1/\beta^2})^2 - 1} - \frac{A_1^2/4\beta^4}{(1-\sqrt{1-1/\beta^2})^2 - 1}$$

$$\left. + \frac{3}{8\beta^2}\left(1 - \frac{6}{\beta^2} + \frac{5}{\beta^4}\right) A_2^2 + \frac{1}{4\beta^2}\left(1 - \frac{3}{\beta^2}\right) A_1^2 \right] A_2 \cos\phi_2$$

$$+ \left[-\frac{3}{8\beta^4}\left(1 - \frac{1}{\beta^2}\right) A_2^2 + \frac{1}{8\beta^2}\left(1 - \frac{6}{\beta^2} + \frac{5}{\beta^4}\right) A_2^2 \right] \times$$

$$\times A_2 \cos 3\phi_2 + \left[\frac{3/8\beta^2(1-1/\beta^2)A_1^2 A_2}{(1-1/\beta^2) - 4} - \frac{A_1^2 A_2/4\beta^2}{(1-\sqrt{1-1/\beta^2})^2 - 1} \right.$$

$$\left. + \frac{1}{8\beta^2}\left(1 - \frac{3}{\beta^2}\right) A_1^2 A_2 \right] \cos(2\phi_1 - \phi_2). \tag{56}$$

From which one obtains

$$\theta_{1_{\xi_2}} = -\frac{1}{2\beta^2}\left[\frac{-1/4\beta^2}{(1-1/\beta^2)} \{3(1-1/\beta^2)A_2^2 + A_1^2\} \right.$$

$$- \frac{A_1^2/8\beta^2}{(1-1/\beta^2) - 4} + \frac{A_2^2/4\beta^2}{(1+\sqrt{1-1/\beta^2})^2 - 1}$$

$$\left. + \frac{A_2^2/4\beta^2}{(1-\sqrt{1-1/\beta^2})^2 - 1} - \frac{3A_1^2}{8} - \frac{1}{4}\left(1 - \frac{3}{\beta^2}\right) A_2^2 \right] \tag{57}$$

$$A_{1_{\xi_2}} = 0 \tag{58}$$

$$\begin{aligned} u_{2_{\xi\xi}} + u_2 = &\left[-\frac{A_1 A_2^2}{8\beta^4} - \frac{A_1 A_2^2/4\beta^4}{(1+\sqrt{1-1/\beta^2})^2 - 1} \right. \\ &\left. + \frac{1}{8\beta^2}\left(1 - \frac{3}{\beta^2}\right) A_1 A_2^2 \right] \cos(\phi_1 + 2\phi_2) \\ &+ \left[-\frac{A_1 A_2^2}{8\beta^4} - \frac{A_1 A_2^2/4\beta^4}{(1-\sqrt{1-1/\beta^2})^2 - 1} \right. \\ &\left. + \frac{1}{8\beta^2}\left(1 - \frac{3}{\beta^2}\right) A_1 A_2^2 \right] \cos(\phi_1 - 2\phi_2) \\ &+ \left[\frac{A_1^3/8\beta^4}{(1-1/\beta^2) - 4} + \frac{A_1^3}{8\beta^2} \right] \cos 3\phi_1 \end{aligned} \tag{59}$$

$$\begin{aligned} \theta_{2_{\xi_2}} = \frac{1}{2\sqrt{1-1/\beta^2}} &\left[\frac{3}{4\beta^4} \left\{ \left(1 - \frac{1}{\beta^2}\right) A_2^2 + A_1^2 \right\} \right. \\ &- \frac{3}{8\beta^4}\left(1 - \frac{1}{\beta^2}\right) A_2^2 - \frac{A_1^2/4\beta^4}{(1+\sqrt{1-1/\beta^2})^2 - 1} \\ &- \frac{A_1^2/4\beta^4}{(1-\sqrt{1-1/\beta^2})^2 - 1} + \frac{3}{8\beta^2}\left(1 - \frac{6}{\beta^2} + \frac{5}{\beta^4}\right) A_2^2 \\ &\left. + \frac{1}{4\beta^2}\left(1 - \frac{3}{\beta^2}\right) A_1^2 \right] \end{aligned} \tag{60}$$

$$A_{2_{\xi_2}} = 0 \tag{61}$$

$$\begin{aligned} v_{2_{\xi\xi}} &+ \left(1 - \frac{1}{\beta^2}\right) v_2 \\ = &\left[-\frac{3}{8\beta^4}\left(1 - \frac{1}{\beta^2}\right) A_2^3 \right. \\ &\left. + \frac{1}{8\beta^2}\left(1 - \frac{6}{\beta^2} + \frac{5}{\beta^4}\right) A_2^3 \right] \cos 3\phi_2 \end{aligned}$$

$$+\left[\frac{3/8\beta^2(1-1/\beta^2)A_1^2A_2}{(1-1/\beta^2)-4}-\frac{A_1^2A_2/4\beta^2}{(1+\sqrt{1-1/\beta^2})^2-1}\right.$$

$$\left.+\frac{1}{8\beta^2}\left(1-\frac{3}{\beta^2}\right)A_1^2A_2\right]\cos(2\phi_1+\phi_2)$$

$$+\left[\frac{3/8\beta^2(1-1/\beta^2)A_1^2A_2}{(1-1/\beta^2)-4}-\frac{A_1^2A_2/4\beta^2}{(1-\sqrt{1-1/\beta^2})^2-1}\right.$$

$$\left.+\frac{1}{8\beta^2}\left(1-\frac{3}{\beta^2}\right)A_1^2A_2\right]\cos(2\phi_1-\phi_2). \tag{62}$$

Equations (58) and (61) imply that A_1 and A_2 are constants to $0(\varepsilon^2)$. Solving equations (59) and (62), one obtains

$$u_2=-\frac{A_1A_2^2}{4\beta^2}\left[-\frac{1}{2\beta^2}-\frac{1/\beta^2}{(1+\sqrt{1-1/\beta^2})^2-1}+\frac{1}{2}(1-3/\beta^2)\right]\times$$

$$\times\frac{\cos(\phi_1+2\phi_2)}{(1+2\sqrt{1-1/\beta^2})^2-1}-\frac{A_1A_2^2}{4\beta^2}\left[-\frac{1}{2\beta^2}\right.$$

$$\left.-\frac{1/\beta^2}{(1-\sqrt{1-1/\beta^2})^2-1}+\frac{1}{2}(1-3/\beta^2)\right]\frac{\cos(\phi_1-2\phi_2)}{(1-2\sqrt{1-1/\beta^2})^2-1}$$

$$-\frac{A_1^3}{64\beta^2}\left[\frac{1/\beta^2}{(1-1/\beta^2)-4}+1\right]\cos 3\phi_1 \tag{63}$$

$$v_2=-\frac{A_1^2A_2}{4\beta^2}\left[\frac{3/4(1-1/\beta^2)}{(1-1/\beta^2)-4}-\frac{1}{(1+\sqrt{1-1/\beta^2})^2-1}\right.$$

$$\left.+\frac{1}{2}\left(1-\frac{3}{\beta^2}\right)\right]\frac{\cos(2\phi_1+\phi_2)}{(2+\sqrt{1-1/\beta^2})^2-(1-1/\beta^2)}$$

$$-\frac{A_1^2A_2}{4\beta^2}\left[\frac{3/4(1-1/\beta^2)}{(1-1/\beta^2)-4}-\frac{1}{(1-\sqrt{1-1/\beta^2})^2-1}\right.$$

$$\left.+\frac{1}{2}\left(1-\frac{3}{\beta^2}\right)\right]\frac{\cos(2\phi_1-\phi_2)}{(2-\sqrt{1-1/\beta^2})^2-(1-1/\beta^2)}$$

$$-\frac{A_2^3}{64\beta^4}\left[-\frac{3}{\beta^2}+\frac{(1-6/\beta^2+5/\beta^4)}{(1-1/\beta^2)}\right]\cos 3\phi_2. \tag{64}$$

Let us now examine these results for some known special cases. First, let us consider a quasi-transverse wave, i.e., $A_2 = 0$ (because then from (38), to $0(\varepsilon)$, $P_z = 0$). (57) gives

$$\theta_{1_{\xi_2}} = -\frac{(9\beta^2 - 1)}{16(\beta^2 - 1)(3\beta^2 + 1)} A_1^2. \tag{65}$$

The wave frequency ω (= uk, k being the wavenumber) is then given on using (45) and (65) by

$$\omega = \phi_{1_t} = \phi_{1_\xi} \cdot \xi_t$$

$$= \frac{\omega_p \omega}{\sqrt{\omega^2 - k^2c^2}} \left[1 - \varepsilon^2 \frac{(9\beta^2 - 1)}{16(\beta^2 - 1)(3\beta^2 + 1)} A_1^2 + \cdots \right]. \tag{66}$$

Using $V_x \approx eE_x/m\omega$, one obtains from (66) an expression showing the amplitude-dependent frequency shift:

$$\omega^2 = \omega_p^2 + k^2c^2 - \varepsilon^2 \left[\frac{e^2\omega_p^2 E_x^2}{m^2c^2\omega^2} \frac{8\omega^2 + \omega_p^2}{8(4\omega^2 - \omega_p^2)} \right] + 0(\varepsilon^3), \tag{67}$$

which is identical to the result of Sluijter and Montgomery [108].

Next, let us consider a quasi-longitudinal wave, i.e., $A_1 = 0$. Then, (60) gives

$$\theta_{2_{\xi_2}} = \frac{3A_2^2/\beta^6}{16\sqrt{1 - 1/\beta^2}} (\beta^4 - \beta^2). \tag{68}$$

The wave frequency ω is then given on using (46) and (68), by

$$\omega = \frac{\omega_p \cdot \beta}{\sqrt{\beta^2 - 1}} \left[\sqrt{1 - 1/\beta^2} - \varepsilon^2 \frac{3}{16} \frac{(\beta^2 - 1)}{\beta^4\sqrt{1 - 1/\beta^2}} A_2^2 + \cdots \right] \tag{69}$$

or

$$\omega = \omega_p \left[1 - \varepsilon^2 \frac{3}{16} \left| \frac{V_z}{c} \right|^2 + 0(\varepsilon^3) \right] \tag{70}$$

which is identical to the one derived by Akhiezer and Polovin [101].

In the above treatments only the electron motion was considered, while the ion motion was ignored. The results obtained thus apply to comparatively small amplitudes for which the ion dynamics is insignificant because of its large rest mass. Nevertheless, in the presence of very large amplitude-waves, ions can attain considerable directed speeds. Therefore, an accurate description of the large-amplitude wave phenomena requires the inclusion of ion dynamics in the analysis (Chian [193]).

(ii) Nonlinear Relativistic Waves in a Magnetized Plasma

Experiments of Stamper *et al.* [109—111] and Diverglio *et al.* [112] showed that intense spontaneously-generated magnetic fields (of the order of a kilo-Gauss) are present in laser-produced plasmas. In the experiment of Stamper *et al.* [109], the target was located at the center of a large tube containing an ambient gas. On focusing the laser beam onto the target, magnetic fields were observed as pulses which propagated with the same velocity as the fronts observed by optical means. Figure 6.2 shows the radial variation of the maximum azimuthal field observed in the midplane ($z = 0$) for a Lucite target in $200 - m$ Torr background of nitrogen. For $r > 1$ cm, $B \sim r^{-1.4}$; in this range, the front was passing the probe while the laser pulse was still incident on the target. For $r > 1$ cm, $B \sim r^{-4.2}$; in this range, the laser pulse was over before the front could reach the probe. Stamper *et al.* [110, 111] made direct observations of spontaneous megagauss magnetic fields in laser-produced plasmas using the Faraday-rotation measurements of plane-polarized electromagnetic radiation traversing the region of high fields. The generation of these

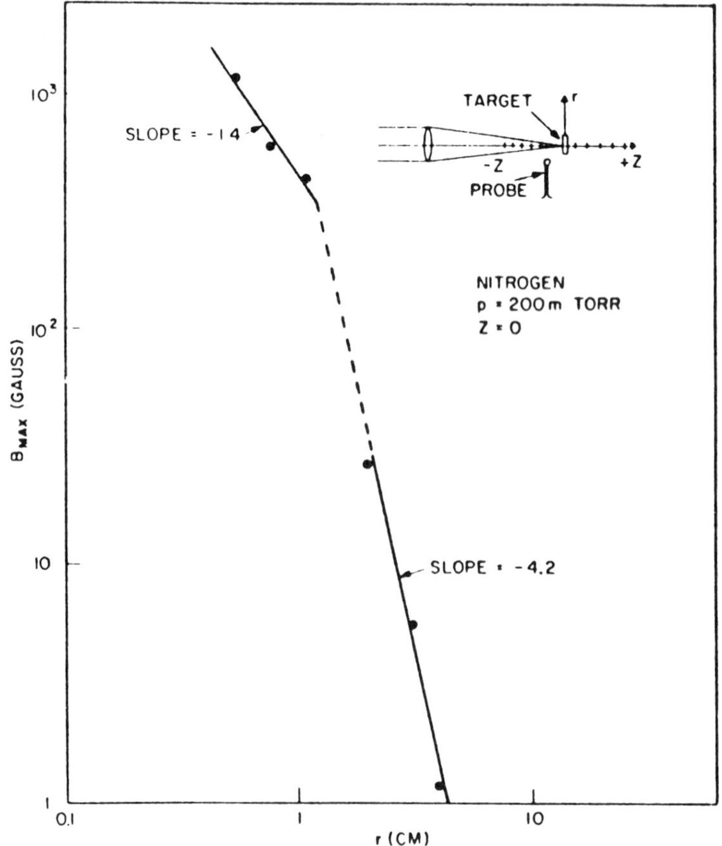

Figure 6.2. Radial variation of spontaneous fields. (Due to Stamper *et al.* [109], by courtesy of The American Physical Society).

magnetic fields was explained in terms of thermoelectric currents associated with large temperature gradients near the target.

Diverglio *et al.* [112] simulated the laser-plasma interaction by the resonant interaction of an electromagnetic wave with an inhomogeneous, collisionless plasma. A quasistatic magnetic field **B** was observed to be generated in the resonance region by the incident electromagnetic wave. The magnetic field **B** was nearly perpendicular to the density gradient along the z-direction and to the incident-wave polarization along the x-direction, i.e., $\mathbf{B} \approx B_y\hat{\mathbf{i}}_y$. Figure 6.3 shows B_x and B_y components in the x, y-plane. Observe that B_y changes sign as a function of x, and falls off monotonically in the y-direction away from $y = 0$.

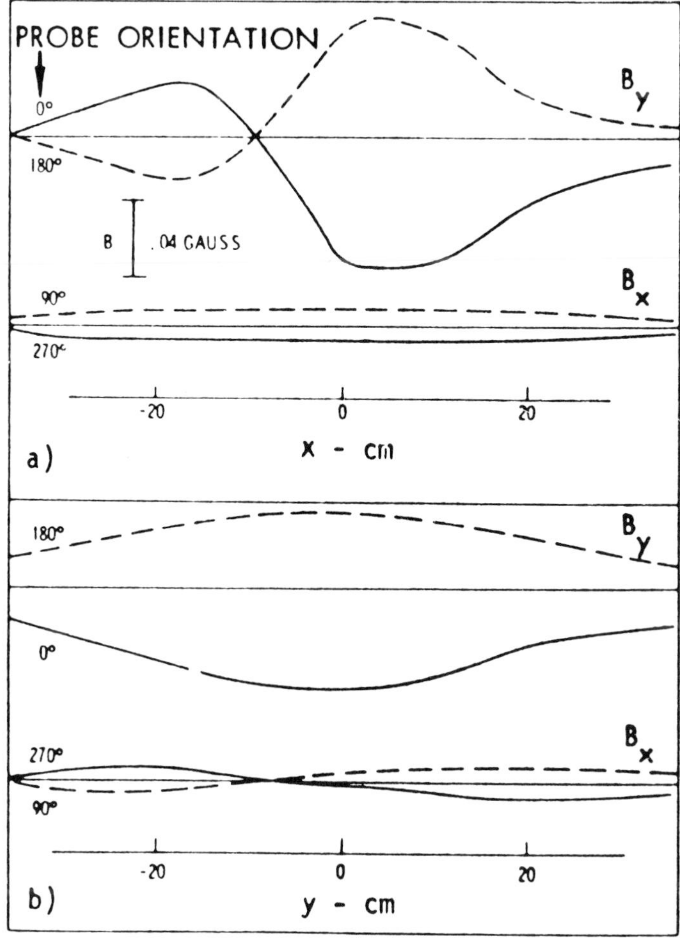

Figure 6.3. Radial components of magnetic field at fixed axial position of maximum field, (a) as a function of x for $y = 10$ cm, (b) as a function of y for $x = 10$ cm. The fixed y and x positions for (a) and (b), respectively, were chosen to be far from lines of $B = 0$. Note the signal sign reversal for probe orientations differing by 180°, indicating the magnetic origin of the signal. (Due to Diverglio *et al.* [112], by courtesy of The American Physical Society).

Because of this sharp reversal of field direction in the resonant region, Diverglio *et al.* [112] argued that resonant absorption of the incident wave was the generating mechanism for magnetic fields.

We will therefore consider now nonlinear relativistic propagation of electromagnetic wave in a magnetized plasma (Akhiezer and Polovin [101], Aliev and Kuznetsov [113], Berzhiani *et al.* [114], and Shivamoggi [115]).

Consider an intense circularly-polarized electromagnetic wave propagating along a constant magentic field $\mathbf{B}_0 = B_0\hat{\mathbf{i}}_z$ in a plasma. Under the action of this wave field, let us assume that the electrons form a cold fluid (i.e., their directed speeds are much larger than their random speeds) undergoing nonlinear oscillations, whereas the ions make up an immobile neutralizing positive-charge background. The basic equations describing the propagation of this circularly-polarized electromagnetic wave coupled nonlinearly to the Langmuir wave in the plasma are

$$\nabla \times \mathbf{E} = -\frac{1}{c}\frac{\partial \mathbf{B}}{\partial t} \tag{71}$$

$$\nabla \times \mathbf{B} = -\frac{4\pi}{c} ne\mathbf{V} + \frac{1}{c}\frac{\partial \mathbf{E}}{\partial t} \tag{72}$$

$$\nabla \cdot \mathbf{E} = -4\pi e(n - n_0) \tag{73}$$

$$\nabla \cdot \mathbf{B} = 0 \tag{74}$$

$$\frac{\partial \mathbf{p}}{\partial t} + (\mathbf{V}\cdot\nabla)\mathbf{p} = -e\left[\mathbf{E} + \frac{1}{c}\mathbf{V}\times(\mathbf{B}_0 + \mathbf{B})\right], \tag{75}$$

where

$$\mathbf{p} = \frac{m\mathbf{V}}{\sqrt{1 - V^2/c^2}}\,. \tag{76}$$

Let us now choose the electromagnetic field for the coupled system with a polarization shown in Figure 6.1. Assuming that all quantities depend only on z and t, we have from equations (71)—(75):

$$\frac{\partial E_x}{\partial z} = -\frac{1}{c}\frac{\partial B_y}{\partial t} \tag{77}$$

$$\frac{\partial E_y}{\partial z} = \frac{1}{c}\frac{\partial B_x}{\partial t} \tag{78}$$

$$\frac{\partial B_x}{\partial z} = -\frac{4\pi}{c} neV_y + \frac{1}{c}\frac{\partial E_y}{\partial t} \tag{79}$$

$$-\frac{\partial B_y}{\partial z} = -\frac{4\pi}{c} neV_x + \frac{1}{c}\frac{\partial E_x}{\partial t} \tag{80}$$

$$\frac{1}{c}\frac{\partial E_z}{\partial t} = \frac{4\pi}{c} neV_z \tag{81}$$

$$\frac{\partial E_z}{\partial z} = -4\pi e(n - n_0) \tag{82}$$

$$\left(\frac{\partial}{\partial t} + V_z \frac{\partial}{\partial z}\right)\left(\frac{mV_x}{\sqrt{1 - V^2/c^2}}\right) = -eE_x + \frac{e}{c}(V_z B_y - V_y B_z) \tag{83}$$

$$\left(\frac{\partial}{\partial t} + V_z \frac{\partial}{\partial z}\right)\left(\frac{mV_y}{\sqrt{1 - V^2/c^2}}\right) = -eE_y + \frac{e}{c}(V_x B_0 - V_z B_x) \tag{84}$$

$$\left(\frac{\partial}{\partial t} + V_z \frac{\partial}{\partial z}\right)\left(\frac{mV_z}{\sqrt{1 - V^2/c^2}}\right) = -eE_z + \frac{e}{c}(V_x B_y - V_y B_x). \tag{85}$$

Let us look for plane nonlinear travelling wave-solutions of (77)—(85), of the form

$$q(z, t) = q(\eta), \quad \eta \equiv z - ut. \tag{86}$$

Then equations (77)—(82) give:

$$\begin{gathered} E_x = \frac{u}{c} B_y, \quad E_y = -\frac{u}{c} B_x, \quad E_z' = -\frac{4\pi n_0 e}{u - V_z} V_z \\ B_x' = \frac{4\pi n_0 e}{c(\beta^2 - 1)} \frac{uV_y}{u - V_z}, \quad B_y' = -\frac{4\pi n_0 e}{c(\beta^2 - 1)} \frac{uV_x}{u - V_z} \\ n = \frac{n_0 u}{u - V_z} \end{gathered} \tag{87}$$

where primes denote differentiation with respect to η, and $\beta \equiv u/c$.

Using (87), and introducing

$$\mathbf{P} = \frac{\mathbf{V}/c}{\sqrt{1 - V^2/c^2}}, \quad \Omega_0 = \frac{eB_0}{mc} \tag{88}$$

one may derive from equations (83)—(85):

$$\begin{aligned} P_x'' &+ \frac{\omega_p^2}{c^2} \frac{\beta P_x}{(\beta^2 - 1)\,[\beta\sqrt{1 + P^2} - P_z]} \\ &- \frac{\Omega_0}{c}\left[\frac{P_y}{\beta\sqrt{1 + P^2} - P_z}\right]' = 0 \end{aligned} \tag{89}$$

$$P_y'' + \frac{\omega_p^2}{c^2} \frac{\beta P_y}{(\beta^2 - 1)[\beta\sqrt{1 + P^2} - P_z]}$$

$$+ \frac{\Omega_0}{c}\left[\frac{P_x}{\beta\sqrt{1 + P^2} - P_z}\right]' = 0 \tag{90}$$

$$(\beta P_z - \sqrt{1 + P^2})'' + \frac{\omega_p^2}{c^2} \frac{P_z}{\beta\sqrt{1 + P^2} - P_z} = 0. \tag{91}$$

Analytical solutions of these equations have not been obtained for the general case of coupled longitudinal-transverse waves. Exact solutions have been obtained only for some special cases like purely transverse waves and purely longitudinal waves.

(a) Purely Transverse Waves

Setting $P_z = 0$, one obtains from equation (91):

$$P^2 = \text{const.} \tag{92}$$

Let us consider a left-hand circularly-polarized wave, so from (92) we have

$$\begin{aligned} P_x &= P\cos\frac{\omega}{\beta}\eta \\ P_y &= P\sin\frac{\omega}{\beta}\eta. \end{aligned} \tag{93}$$

Using (93), equations (89) and (90) give

$$\omega\left[\omega + \frac{\Omega_0}{\sqrt{1 + P^2}}\right] - \frac{\beta^2\omega_p^2}{(\beta^2 - 1)} \frac{1}{\sqrt{1 + P^2}} = 0. \tag{94}$$

Using (94), one may write

$$u = \beta c = \frac{c}{\sqrt{\varepsilon}} \tag{95}$$

where,

$$\varepsilon = 1 - \frac{\omega_p^2/\omega^2}{\sqrt{1 + P^2} + \Omega_0/\omega}. \tag{96}$$

Now, from equations (77)—(85), one has

$$\begin{aligned} E_x &= -\frac{mc\omega P}{e}\left[1 + \frac{\Omega_0/\omega}{\sqrt{1 + P^2}}\right]\sin\frac{\omega}{\beta}\eta \\ E_y &= \frac{mc\omega P}{e}\left[1 + \frac{\Omega_0/\omega}{\sqrt{1 + P^2}}\right]\cos\frac{\omega}{\beta}\eta \end{aligned} \tag{97}$$

from which,

$$E_0^2 = E_x^2 + E_y^2$$
$$= \frac{m^2c^2\omega^2P^2}{e^2}\left[1 + \frac{\Omega_0/\omega}{\sqrt{1+P^2}}\right]^2. \tag{98}$$

In order to invert this, let us assume that $\Omega_0/\omega \ll 1$. We then obtain

$$P^2 \approx \frac{e^2E_0^2}{m^2c^2\omega^2}\left[1 + \frac{\Omega_0/\omega}{\sqrt{1+e^2E_0^2/m^2c^2\omega^2}}\right]^{-2}. \tag{99}$$

Using (99) in (96), and recalling that wave propagation is possible only when $\varepsilon > 0$, one finds that waves in the frequency range

$$\omega_p\left[1 + \frac{e^2E_0^2}{m^2c^2\omega^2}\left\{1 + \frac{\Omega_0}{\omega}\left(1 + \frac{e^2E_0^2}{m^2c^2\omega^2}\right)^{-1/2}\right\}^{-2}\right]^{-1/2} < \omega < \omega_p \tag{100}$$

can also propagate in the plasma. Observe that according to (100), the frequency range for wave propagation is widened further in the presence of a constant applied magnetic field $\mathbf{B}_0$.

(b) *Waves with* $\beta \cong 1$

In order to treat waves with $\beta \cong 1$, let us introduce as before

$$\theta = \frac{(\omega_p/c)}{\sqrt{\beta^2-1}}\,\eta \tag{101}$$

then equations (89)—(91) give:

$$\frac{d^2P_x}{d\theta^2} + P_x[\sqrt{1+P^2} - P_z] = 0 \tag{102}$$

$$\frac{d^2P_y}{d\theta^2} + P_y[\sqrt{1+P^2} - P_z] = 0 \tag{103}$$

$$\frac{d^2}{d\theta^2}[P_z - \sqrt{1+P^2}] + (\beta^2-1)P_z[\sqrt{1+P^2} - P_z] = 0 \tag{104}$$

which are identical to those equations, namely (28)—(30), for the case of an unmagnetized plasma. Therefore, the constant applied magnetic field $\mathbf{B}_0$ does not affect the waves for which $\beta \cong 1$.

(iii) Relativistic Modulational Instability of an Electromagnetic Wave

The relativistic mass variation of the electrons moving in an intense electromagnetic wave can also lead to a modulational instability of the latter (Max *et al.* [116]). This instability mechanism involves only electrons because the instability can grow so quickly in time that ion inertia prevents the ions from following along. We consider here the modulational instability of an intense electromagnetic wave in a plasma using a method given by Shivamoggi [117].

Consider the propagation of an intense linearly-polarized electromagnetic wave of frequency ω_0 and wavenumber k_0 in a plasma. In the modulation interaction problem, the plasma response is two-pronged: the high-frequency motion of the electrons in the wave field, and the low-frequency motion caused by the ponderomotive force exerted by the self-interaction of the wave. The relativistic modulational processes that occur in cases involving intense electromagnetic waves occur at a rate rapid for the ions to participate in them. Therefore, we consider the ions to remain immobile and to form a neutralizing background, and consider only the electron response. Further, we shall neglect the thermal motion of the electrons in comparison with the directed motion. We then have the following equations governing the motion of the electron fluid:

$$\frac{\partial n}{\partial t} + \nabla \cdot (n\mathbf{V}) = 0 \tag{105}$$

$$\frac{\partial \mathbf{p}}{\partial t} + (\mathbf{V} \cdot \nabla)\mathbf{p} = -e\left[\mathbf{E} + \frac{1}{c}\mathbf{V} \times \mathbf{B}\right] \tag{106}$$

$$\nabla \times \mathbf{E} = -\frac{1}{c}\frac{\partial \mathbf{B}}{\partial t} \tag{107}$$

$$\nabla \times \mathbf{B} = -\frac{4\pi}{c}ne\mathbf{V} + \frac{1}{c}\frac{\partial \mathbf{E}}{\partial t} \tag{108}$$

$$\nabla \cdot \mathbf{E} = -4\pi e(n - n_0) \tag{109}$$

$$\nabla \cdot \mathbf{B} = 0 \tag{110}$$

where

$$\mathbf{p} = \frac{m\mathbf{V}}{\sqrt{1 - V^2/c^2}}. \tag{111}$$

Let us introduce the scalar and vector potentials ϕ and $\mathbf{A}$ to describe the wave fields $\mathbf{E}$ and $\mathbf{B}$:

$$\mathbf{B} = \nabla \times \mathbf{A}, \quad \mathbf{E} = -\nabla\phi - \frac{1}{c}\frac{\partial \mathbf{A}}{\partial t}, \tag{112}$$

ϕ turns out to be the low-frequency ambipolar potential.

Let us consider an electromagnetic wave given by

$$\mathbf{A} = A(z, t)\cos\theta\hat{\mathbf{i}}_x, \quad \theta = k_0 z - \omega_0 t \tag{113}$$

to propagate along the z-axis, with the linear wave propagation relation

$$\omega_0^2 = k_0^2 c^2 + \omega_p^2. \tag{114}$$

Using (111) and (112), equation (106) can be written as

$$\frac{\partial \mathbf{p}}{\partial t} + \frac{1/m}{\sqrt{1 + p^2/m^2c^2}} (\mathbf{p} \cdot \nabla)\mathbf{p} = \frac{e}{c} \frac{\partial \mathbf{A}}{\partial t} + e\nabla\phi$$

$$- \frac{e/mc}{\sqrt{1 + p^2/m^2c^2}} [\mathbf{p} \times (\nabla \times \mathbf{A})] \tag{115}$$

from which one has

$$A \cos \theta = \frac{cp}{e} \tag{116}$$

$$\nabla\phi - c^2\nabla \left(\frac{1}{\sqrt{1 + p^2/m^2c^2}} \right) = 0. \tag{117}$$

Note that equation (117) has been obtained by ignoring the electron momentum in the wave propagation direction in the low-frequency plasma response.

Next, one obtains from equations (107), (108) and (110), the wave equation

$$\frac{\partial^2 \mathbf{A}}{\partial t^2} - c^2\nabla^2\mathbf{A} = -4\pi nec\mathbf{V}. \tag{118}$$

Using (113) and (114), and noting that $A(z, t)$ is a slowly-varying function of z and t, one obtains from equation (118)

$$\left[i \left(\frac{\partial A}{\partial t} + \frac{k_0 c^2}{\omega_0} \frac{\partial A}{\partial z} \right) + \frac{c^2}{2\omega_0} \frac{\partial^2 A}{\partial z^2} \right] \cos \theta$$

$$= \frac{\frac{\omega_p^2}{2\omega_0} \frac{cp}{e} \frac{n}{n_0}}{\sqrt{1 + p^2/m^2c^2}} \tag{119}$$

where,

$$n = n_0 + \delta n. \tag{120}$$

The density perturbation δn can be calculated from the dynamics of the plasma motion parallel to the wave propagation direction

$$\frac{\partial \delta n}{\partial t} + n_0 \frac{\partial \delta V}{\partial z} = 0 \tag{121}$$

$$\frac{\partial \delta V}{\partial t} = -\frac{e}{m} \left(-\frac{1}{c} V_0 \, \delta B + \delta E_z \right) \tag{122}$$

$$\frac{\partial \delta E}{\partial z} = -4\pi e \, \delta n \tag{123}$$

where,

$$V_0 = -\frac{eE_0}{im\omega_0} e^{i\theta}$$
$$\delta B = \frac{ik_0c}{\omega_0} E_0 e^{i\theta}. \tag{124}$$

One obtains from equations (121)—(124):

$$\frac{\delta n}{n_0} = -\frac{e^2E_0^2}{m^2c^2\omega_0^2}\frac{\omega_0^2-\omega_p^2}{4\omega_0^2-\omega_p^2}\cos 2\theta$$

or

$$\frac{\delta n}{n_0} = \frac{e^2A^2}{m^2c^4}\frac{\omega_0^2-\omega_p^2}{4\omega_0^2-\omega_p^2}\cos 2\theta. \tag{125}$$

Next, one obtains from (116),

$$p = \frac{eA}{c}\cos\theta. \tag{126}$$

Using (120), (125) and (126), equation (119) becomes

$$i\left(\frac{\partial A}{\partial t} + \frac{k_0^2c^2}{\omega_0}\frac{\partial A}{\partial z}\right) + \frac{c^2}{2\omega_0}\frac{\partial^2 A}{\partial z^2} + \frac{\omega_p^2 q}{4\omega_0}|A|^2A = 0 \tag{127}$$

which is the familiar nonlinear Schrödinger equation. Here,

$$q \equiv \frac{e^2}{m^2c^4}\left(\frac{3}{4} - \frac{\omega_0^2-\omega_p^2}{4\omega_0^2-\omega_p^2}\right). \tag{128}$$

Equation (127) can be seen to arise through the familiar nonlinear frequency shift due to the relativistic effects, namely, (67).

Upon transferring to a frame moving with the group velocity k_0c^2/ω_0 of the wave, equation (127) reduces to the standard form

$$i\frac{\partial A}{\partial t} + \frac{c^2}{2\omega_0}\frac{\partial^2 A}{\partial \xi^2} + \frac{\omega_p^2 q}{4\omega_0}|A|^2A = 0 \tag{129}$$

where,

$$\xi \equiv z - \left(\frac{k_0c^2}{\omega_0}\right)t.$$

In order to investigate stability with respect to low-frequency modulations, let us rewrite equation (129) as follows:

$$i\frac{\partial A}{\partial t} + \frac{c^2}{2\omega_0}\frac{\partial^2 A}{\partial \xi^2} + \frac{\omega_p^2 q}{4\omega_0}(|A|^2 - |A_0|^2)A = 0 \tag{130}$$

where A is the amplitude on superposing a modulation over the wave with amplitude A_0. On putting,

$$A = \sqrt{P(\xi, t)}\, e^{i\sigma(\xi, t)} \tag{131}$$

equation (130) gives

$$\frac{\partial P}{\partial t} + \frac{c^2}{\omega_0} \frac{\partial}{\partial \xi} \left(P \frac{\partial \sigma}{\partial \xi} \right) = 0 \tag{132}$$

$$\frac{\omega_p^2 q}{4\omega_0} (P - P_0) - \frac{\partial \sigma}{\partial t} + \frac{c^2}{4\omega_0 P} \frac{\partial^2 P}{\partial \xi^2} = 0. \tag{133}$$

On putting further

$$P = P_0 + P_1(\xi, t), \quad \sigma = \sigma_1(\xi, t) \tag{134}$$

linearizing in P_1 and σ_1, and looking for solutions of the form

$$P_1(\xi, t), \quad \sigma_1(\xi, t) \sim e^{i(K\xi - \Omega t)} \tag{135}$$

one obtains from equations (132) and (133):

$$-i\Omega P_1 - \frac{P_0 c^2}{\omega_0} K^2 \sigma_1 = 0 \tag{136}$$

$$\left(\frac{\omega_p^2 q}{4\omega_0} - \frac{c^2 K^2}{4\omega_0 P_0} \right) P_1 + i\Omega \sigma_1 = 0 \tag{137}$$

from which one has

$$\Omega^2 = \frac{c^2 K^2}{4\omega_0} \left(\frac{c^2 K^2}{\omega_0} - \frac{\omega_p^2 q}{\omega_0} |A_0|^2 \right). \tag{138}$$

According to (138), the modulational instability arises if

$$|E_0|^2 > \frac{c^4 K^2 \omega_0^2}{\omega_p^2 q}. \tag{139}$$

The maximum growth rate is given by

$$\Omega_{\max} = \frac{\omega_p^2 c^2 q}{4\omega_0^3} |E_0|^2. \tag{140}$$

VII

Nonlinear Waves in an Inhomogeneous Plasma

The foregoing accounts were concerned with nonlinear waves in a homogeneous plasma. It may be noted that nonlinear waves get modified in an essential way in an inhomogeneous plasma. For instance, the amplitude, width, and propagation velocity of an ion-acoustic solitary wave will change as the latter moves in an inhomogeneous plasma (Nishikawa and Kaw [194], John and Saxena [195], Rao and Varma [196], Shivamoggi [197, 198], Kuehl [199, 200], Chang *et al.* [201]). Another essential effect of plasma inhomogeneity is that often the instabilities become convective and the matching of selection rules for resonant wave-wave interactions is destroyed in a distance short compared with effective growth lengths (Chakraborty [202, 203], Rosenbluth *et al.* [204, 205], Pesme *et al.* [206], Liu *et al.* [207], Lee and Kaw [208], Pramanik [209], Grebogi and Liu [210], Lindegren *et al.* [211]). Here, we will give a brief discussion of the propagation characteristics of ion-acoustic solitary waves in an inhomogeneous plasma. For the other effects of plasma inhomogeneity on nonlinear waves, we refer the reader to the original literature cited above.

(i) Ion-acoustic Solitary Waves in an Inhomogeneous Plasma

Theoretical treatments describing the properties of an ion-acoustic solitary wave in an inhomogeneous plasma have been based on Korteweg—deVries equation which is modified either by variable coefficients or by additional small terms. These theories give, for example, the spatial dependence of the slowly-varying soliton-amplitude, width, and speed in an inhomogeneous plasma. The soliton behavior is determined by carrying out a perturbation expansion based on the assumption that the soliton width is small compared with the scale length of the plasma inhomogeneity. Under this condition the soliton retains its identity, and its amplitude, width and speed are slowly-varying functions of position. Early theoretical treatments on this problem were either physically inconsistent due to the neglect of the ion-drift in the unperturbed state (Nishikawa and Kaw [195] and Shivamoggi [197]) or physically unrealizable due to the neglect of an ion-generation without which it is impossible to have an ion-drift in the unperturbed state (Rao and Varma [196]). Fully satisfactory treatments have been recently given by Kuehl and Imen [199, 200] and Shivamoggi [198] which include the ion-drift in conjunction with an ion-generation in a self-consistent way.

Consider an ion-acoustic solitary wave propagating in an inhomogeneous plasma which is otherwise in a time-independent state. Assuming that the ions are cold and that the electrons follow a Boltzmann distribution, we have for the ions:

$$\frac{\partial n}{\partial t} + \frac{\partial}{\partial x}(nv) = \nu_1 \tag{1}$$

$$\frac{\partial v}{\partial t} + v\frac{\partial v}{\partial x} = -\frac{\partial \phi}{\partial x} - \nu_1 \frac{v}{n} \tag{2}$$

$$\frac{\partial^2 \phi}{\partial x^2} = n_0 e^{\phi} - n_0 \tag{3}$$

where n is normalized by a reference value of the unperturbed number density, v is normalized by the ion-acoustic speed C_s, ϕ is normalized by kT_e/e, and t^{-1} is normalized by the ion-plasma frequency ω_{p_i}. ν_1 is the ionization rate per unit volume.

Let us introduce two new independent variables:

$$\xi = \varepsilon^{1/2}\left[\int^x \frac{dx'}{\lambda_0(x')} - t\right], \quad \eta = \varepsilon^{3/2}x \tag{4}$$

where ε is a small parameter characterizing the typical amplitude of a wave, and seek solutions of the form:

$$\begin{aligned} n &= n_0 + \varepsilon n_1 + \varepsilon^2 n_2 + \cdots \\ \phi &= \phi_0 + \varepsilon\phi_1 + \varepsilon^2\phi_2 + \cdots \\ v &= v_0 + \varepsilon v_1 + \varepsilon^2 v_2 + \cdots \\ \nu_1 &= \varepsilon^{3/2}\nu. \end{aligned} \tag{5}$$

Since n_0 and λ_0 (to be determined below) are to be independent of t, we have

$$\frac{\partial n_0}{\partial \xi} = 0, \qquad \frac{\partial \lambda_0}{\partial \xi} = 0. \tag{6}$$

Using (4)—(6), equations (1)—(3) give to $0(1)$:

$$\frac{\partial v_0}{\partial \xi} = 0, \quad \frac{\partial \phi_0}{\partial \xi} = 0. \tag{7}$$

(6) and (7) imply that the unperturbed quantities (denoted by the subscript 0) depend only on the slow-space variable η. These are governed by:

$$\frac{\partial}{\partial \eta}(n_0 v_0) = \nu \tag{8}$$

$$\frac{\partial}{\partial \eta}\left(\frac{1}{2} v_0^2 + \phi_0\right) = -\nu \frac{v_0}{n_0} \tag{9}$$

$$n_0 e^{\phi_0} - n_0 = 0. \tag{10}$$

We have, from (10),

$$\phi_0 \equiv 0. \tag{11}$$

Using (11), (8) and (9) give:

$$n_0 \frac{\partial v_0}{\partial \eta} = -\nu \tag{12}$$

$$v_0 \frac{\partial n_0}{\partial \eta} = 2\nu. \tag{13}$$

Equations (12) and (13) give:

$$n_0 = \frac{1}{B}(\nu\eta + A)^2 \tag{14}$$

$$v_0 = \frac{B}{(\nu\eta + A)}. \tag{15}$$

A and B are constants to be determined using prescribed boundary conditions.

To $0(\varepsilon)$, equations (1)—(3) give:

$$-\frac{\partial n_1}{\partial \xi} + \frac{1}{\lambda_0}\frac{\partial}{\partial \xi}(n_0 v_1 + n_1 v_0) = 0 \tag{16}$$

$$-\frac{\partial v_1}{\partial \xi} + \frac{v_0}{\lambda_0}\frac{\partial v_1}{\partial \xi} + \frac{1}{\lambda_0}\frac{\partial \phi_1}{\partial \xi} = 0 \tag{17}$$

$$-n_0\phi_1 + n_1 = 0 \tag{18}$$

from which, using (6) and (7), we may derive:

$$\frac{n_0}{\lambda_0}\left[1 - \lambda_0^2\left(1 - \frac{v_0}{\lambda_0}\right)^2\right]\frac{\partial v_1}{\partial \xi} = 0 \tag{19}$$

from which,

$$1 - \lambda_0^2\left(1 - \frac{v_0}{\lambda_0}\right)^2 = 0$$

or

$$\lambda_0 = 1 + v_0. \tag{20}$$

Using (20), equation (16) then gives:

$$n_1 = n_0 v_1 \tag{21}$$

Next, to $0(\varepsilon^2)$, equations (1)—(3) give:

$$-\frac{\partial n_2}{\partial \xi} + \frac{1}{\lambda_0}\frac{\partial}{\partial \xi}(n_0 v_2 + n_2 v_0 + n_1 v_1) + \frac{\partial}{\partial \eta}(n_0 v_1 + n_1 v_0) = 0 \tag{22}$$

$$-\frac{\partial v_2}{\partial \xi} + \frac{1}{\lambda_0}\left(v_0 \frac{\partial v_2}{\partial \xi} + v_1 \frac{\partial v_1}{\partial \xi}\right) + v_0 \frac{\partial v_1}{\partial \eta} + v_1 \frac{\partial v_0}{\partial \eta}$$

$$+ \frac{1}{\lambda_0}\frac{\partial \phi_2}{\partial \xi} + \frac{\partial \phi_1}{\partial \eta} = -\nu\phi_1\left(\frac{1 - v_0}{n_0}\right) \tag{23}$$

$$\frac{1}{\lambda_0^2}\frac{\partial^2 \phi_1}{\partial \xi^2} - n_0\phi_2 - \frac{1}{2} n_0 \phi_1^2 + n_2 = 0. \tag{24}$$

Using (6)—(13), (18), (20) and (21), we may derive from equations (22)—(24):

$$\frac{\partial \phi_1}{\partial \eta} + \frac{1}{\lambda_0^2}\phi_1 \frac{\partial \phi_1}{\partial \xi} + \left(\frac{1}{2\lambda_0 n_0}\frac{\partial n_0}{\partial \eta}\right)\phi_1 + \frac{1}{2\lambda_0^4 n_0}\frac{\partial^3 \phi_1}{\partial \xi^3} = 0. \tag{25}$$

Putting,

$$\phi_1 = g(\eta)\psi_1, g(\eta) = e^{-\nu \int \frac{1}{2n_0\lambda_0}\left[\frac{2}{v_0} + (1 - v_0)\right] d\eta} \tag{26}$$

equation (25) gives:

$$\frac{\partial \psi_1}{\partial \eta} + \frac{g}{(1 + v_0)^2}\psi_1 \frac{\partial \psi_1}{\partial \xi} + \frac{1}{2(1 + v_0)^4 n_0}\frac{\partial^3 \psi_1}{\partial \xi^3} = 0. \tag{27}$$

In order to determine the properties of the solitary wave governed by equation (27), consider the equation (Ko and Kuehl [218])

$$\frac{\partial \psi_1}{\partial \eta} + \alpha(\eta)\psi_1 \frac{\partial \psi_1}{\partial \xi} + \beta(\eta)\frac{\partial^3 \psi_1}{\partial \xi^3} = 0. \tag{28}$$

To lowest order in ε, equation (28) has the solution

$$\psi_1 \approx a \operatorname{sech}^2 \varphi \tag{29}$$

where,

$$\begin{aligned} &\varphi = b\theta, \quad a = \frac{3\omega}{\alpha k}, \quad b = \sqrt{\frac{\omega}{4\beta k^3}} \\ &k(\eta) = \frac{\partial \theta}{\partial \xi}, \quad \omega(\eta) = -\frac{\partial \theta}{\partial \eta}. \end{aligned} \tag{30}$$

From,

$$\frac{\partial k}{\partial \eta} = -\frac{\partial \omega}{\partial \xi} = 0 \tag{31}$$

we have,

$$k = \text{constant} \tag{32}$$

Equation (28) has an integral invariant to lowest order in ε given by

$$\frac{\partial}{\partial \eta} \int_{-\infty}^{\infty} \psi_1^2 \, \mathrm{d}\theta = 0. \tag{33}$$

Using (29) and (30), (33) gives

$$\frac{\partial}{\partial \eta} \left(\frac{a^2}{b} \right) = 0$$

or

$$\omega = \omega_0 \left(\frac{\alpha}{\alpha_0} \right)^{4/3} \left(\frac{\beta}{\beta_0} \right)^{-1/3} \tag{34}$$

where the subscript 0 refers to some reference values. Using (34), (30) gives

$$a \sim \left(\frac{\alpha}{\beta} \right)^{1/3}. \tag{35}$$

Returning to equation (27), the amplitude of the solitary wave is then given by

$$\psi_{1_{\max}} \sim \left(\frac{g/\lambda_0^2}{1/2\lambda_0^4 n_0} \right)^{1/3}. \tag{36}$$

Thus, from (18) and (26), we have

$$n_{1_{\max}} \sim \mathrm{n}_0 g (2\lambda_0^2 n_0 g)^{1/3}$$

which gives, on using (14), (15), (20) and (26)

$$n_{1_{\max}} \sim n_0 (\sqrt{n_0} + \sqrt{B})^{2/3} e^{-\frac{\nu}{3} \int \frac{2\sqrt{\frac{B}{n_0}} + \frac{B}{n_0}\left(1 - \sqrt{\frac{B}{n_0}}\right)}{2B\left(1 + \sqrt{\frac{B}{n_0}}\right)} \mathrm{d}\eta}$$

or

$$n_{1_{\max}} \sim n_0 (\sqrt{n_0} + \sqrt{B})^{2/3}. \tag{37}$$

On the other hand, the speed c of the solitary wave is given approximately from (20) to be

$$c \approx \lambda_0 = 1 + \sqrt{\frac{B}{n_0}}. \tag{38}$$

(37) and (38) show that the amplitude of the solitary wave increases and the speed decreases as the solitary wave propagates into regions of increasing density. This density-dependence of both amplitude and speed was observed in the early experiment of John and Saxena [194]. This experiment was done in a large double plasma machine. The density scale length was of the order of 20 cm which was many times the solitary width (typically 4 cm). The propagation characteristics of solitary waves in an inhomogeneous plasma are shown in Figures 7.1 and 7.2. Figure 7.1 shows the normalized solitary-wave amplitude plotted against $[n_0(x)/N_0]^{-1/2}$. Figure 7.2 shows the measured solitary-wave velocity plotted against $[n_0(x)/N_0]^{-1/2}$. Here, N_0 is a reference value for the

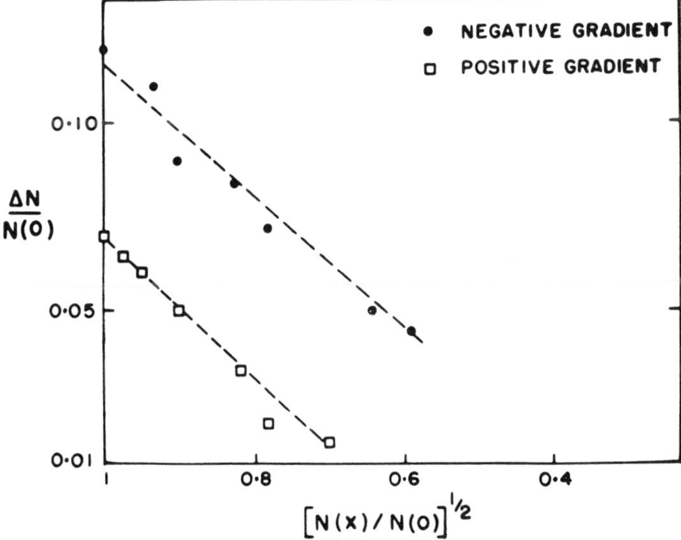

Figure 7.1. Normalized soliton amplitude vs. $[n(x)]^{-1/2}$. (Due to John and Saxena [195], by courtesy of North-Holland Physics Publishing.)

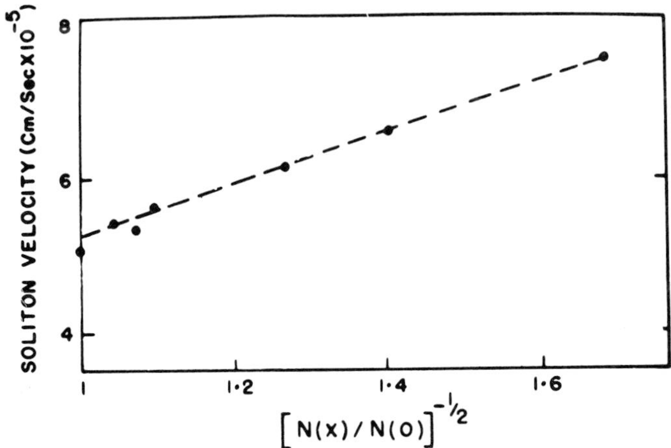

Figure 7.2. Normalized soliton velocity vs. $[n(x)]^{-1/2}$. (Due to John and Saxena [195], by courtesy of North-Holland Physics Publishing.)

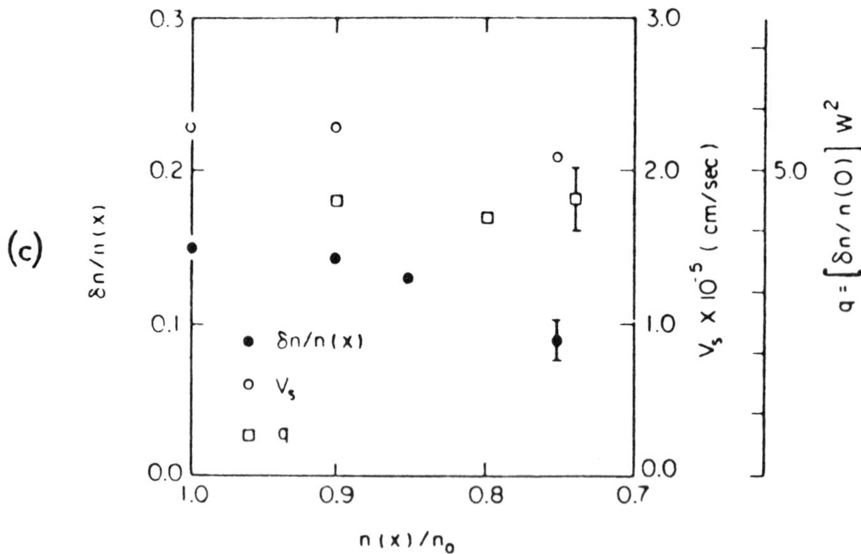

Figure 7.3. Measured soliton amplitude, local velocity, and soliton parameter q as a function of plasma density. (Due to Chang *et al.* [201], by courtesy of The American Institute of Physics.)

number density. The recent experiment of Chang *et al.* [201] also confirmed the above density-dependence of the amplitude, but reported a density-dependence of the speed that is at variance with the above theoretical prediction. This experiment was performed in a multi-dipole-double-plasma device. Chang *et al.* [201] indeed observed a dc drift of particles produced by a density inhomogeneity in the unperturbed state. Figure 7.3 shows that the local amplitude of the solitary wave decreases as the density decreases. The local velocity of the soliton (measured by compensating for the above dc drift velocity) as it propagates into the region of decreasing density tends to decrease. The width appears to adjust itself to this amplitude decrease such that the product q = (amplitude) × (width)^2 remains approximately a constant.

Appendix 1: The Fluid Model for a Plasma

Many features of plasma waves can be described by the fluid model. This simple mathematical model treats the plasma as a mixture of electron-fluid and ion-fluid acted on by electromagnetic and pressure fields. The electron- and ion-fluids are then described in terms of local properties such as the number density and temperature, where the variables are defined as averages over fluid elements that are large compared with the microscopic dimensions of the plasma, such as the Debye length and particle gyroradius, but small in comparison with the macroscopic dimensions. The practical significance of the fluid model is not that it is a very good approximation to any real plasma (indeed, the discrete particle effects are totally outside a fluid model), but rather that it offers the easiest model by which the macroscopic interaction between plasmas and electromagnetic fields may be studied.

Basically, the fluid approximation is valid if there is sufficient localization of the particles in physical space. This localization can be accomplished by two means:

(i) collisions among the particles,
(ii) gyrations of the particles in a strong magnetic field.

In the latter case, the localization is possible only across the field lines.

In order to ensure the validity of the fluid model, one would require that the collisional processes by which the particle distribution function is made isotropic must be rapid compared with the typical time scale in the flow. Next, the electron gyroperiod has to be large compared with the electron mean free time so that the electrons do not move freely in a magnetic field.

The kinetic equation for a collision-dominated plasma is

$$\frac{\partial f_\alpha}{\partial t} + v_k \frac{\partial f_\alpha}{\partial x_k} + a_{\alpha_k} \frac{\partial f_\alpha}{\partial v_k} = \left(\frac{\partial f_\alpha}{\partial t} \right)_c \tag{1}$$

where f_α $\mathrm{d}\mathbf{x}$ $\mathrm{d}\mathbf{V}$ represents the number of particles α in the volume element $\mathrm{d}\mathbf{x}$ $\mathrm{d}\mathbf{V}$ in phase space, and $(\partial f_\alpha/\partial t)_c$ represents the effect due to collisions among the particles, and

$$\mathbf{a}_\alpha = \frac{e_\alpha}{m_\alpha}\left(\mathbf{E} + \frac{1}{c}\,\mathbf{V}\times\mathbf{B}\right).$$

Multiply equation (1) by 1, mv_i, $\frac{1}{2}mv_i^2$ in turn, and integrate over the velocity space. Noting that the number density, momentum and energy are conserved during the collisions, one obtains:

$$\frac{\partial n_\alpha}{\partial t} + \frac{\partial}{\partial x_k}(n_\alpha u_{\alpha_k}) = 0 \tag{2}$$

$$m_\alpha n_\alpha\left(\frac{\partial}{\partial t} + u_{\alpha_k}\frac{\partial}{\partial x_k}\right)u_{\alpha_i} + \frac{\partial p_{\alpha_{ik}}}{\partial x_k} - m_\alpha n_\alpha a_{\alpha_i} = 0 \tag{3}$$

$$\left(\frac{\partial}{\partial t} + u_{\alpha_k}\frac{\partial}{\partial x_k}\right)\left(\frac{3}{2}\,n_\alpha K T_\alpha\right) + \frac{\partial q_{\alpha k}}{\partial x_k} + \frac{\partial u_{\alpha_j}}{\partial x_k}\,p_{\alpha_{jk}}$$

$$+\frac{3}{2}\,n_\alpha K T_\alpha\,\frac{\partial u_{\alpha_k}}{\partial x_k} = 0 \tag{4}$$

where,

$$n_\alpha = \int f_\alpha\,\mathrm{d}\mathbf{V} \tag{5}$$

$$u_{\alpha_i} = \frac{1}{n_\alpha}\int v_i f_\alpha\,\mathrm{d}\mathbf{V} \tag{6}$$

$$p_{\alpha_{ij}} = m_\alpha\int (v_i - u_{\alpha_i})(v_j - u_{\alpha_j})f_\alpha\,\mathrm{d}\mathbf{V} \tag{7a}$$

$$Q_{\alpha_{ijk}} = m_\alpha\int (v_i - u_{\alpha_i})(v_j - u_{\alpha_j})(v_k - u_{\alpha_k})f_\alpha\,\mathrm{d}\mathbf{V} \tag{7b}$$

$$T_\alpha \equiv \frac{p_{\alpha_{ii}}}{3n_\alpha K},\quad q_{\alpha_i} \equiv \frac{1}{2}\,Q_{\alpha_{ijj}}. \tag{7c}$$

For the collision-dominated case, one has to lowest order, from equation (1):

$$\left(\frac{\partial f_\alpha^{(0)}}{\partial t}\right)_c = 0 \tag{8}$$

which, for the Boltzmann collision term, has a solution:

$$f_\alpha^{(0)} = n_\alpha \left(\frac{m_\alpha}{2\pi K T_\alpha} \right)^{3/2} \exp \left[- \frac{m_\alpha (\mathbf{V} - \mathbf{u}_\alpha)^2}{2 K T_\alpha} \right]. \tag{9}$$

Then, the pressure tensor $p_{\alpha_{ij}}$ becomes isotropic:

$$p_{ij}^{(0)} = p_\alpha \delta_{ij}, \quad p_\alpha = n_\alpha K T_\alpha \tag{10}$$

and the heat-flux vector $\mathbf{q}_\alpha$ vanishes:

$$\mathbf{q}_\alpha \equiv 0. \tag{11}$$

Equations (2)—(4), then become:

$$\frac{\partial n_\alpha}{\partial t} + \frac{\partial}{\partial x_k} (n_\alpha u_{\alpha_k}) = 0 \tag{12}$$

$$m_\alpha n_\alpha \left(\frac{\partial}{\partial t} + u_{\alpha_k} \frac{\partial}{\partial x_k} \right) u_{\alpha_i} + \frac{\partial p_\alpha}{\partial x_i} - m_\alpha n_\alpha a_{\alpha_i} = 0 \tag{13}$$

$$\left(\frac{\partial}{\partial t} + u_{\alpha_k} \frac{\partial}{\partial x_k} \right) \left(\frac{3}{2} n_\alpha K T_\alpha \right) + \frac{5}{2} n_\alpha K T_\alpha \frac{\partial u_{\alpha_j}}{\partial x_j} = 0. \tag{14}$$

Using (10) and (12), equation (14) becomes

$$\left(\frac{\partial}{\partial t} + u_{\alpha_k} \frac{\partial}{\partial x_k} \right) (p_\alpha n_\alpha^{-5/3}) = 0 \tag{15}$$

which represents an adiabatic equation of state.

Appendix 2: Review of Linear Waves

(i) Plasma Oscillations

If the electrons in a plasma are displaced from a uniform background of ions, electric fields will be built up in such a direction as to restore the neutrality of the plasma by pulling the electrons back to their original positions. Because of their inertia, the electrons will overshoot and oscillate around their equilibrium positions with a characteristic frequency known as the plasma frequency. This oscillation is so fast that the massive ions do not have time to respond to the oscillating field and may be considered as fixed.

Consider electron oscillations in a uniform plasma, neglecting first the thermal motion of the particles. The ions are assumed to form a uniform fluid providing the static neutralizing background for the electron fluid in equilibrium. The perturbations in the number density obeys

$$\frac{\partial n_1}{\partial t} + n_0 \nabla \cdot \mathbf{V}_1 = 0. \tag{1}$$

The equation of motion gives (in the usual notation)

$$m_e \frac{\partial \mathbf{V}_1}{\partial t} = -e\mathbf{E}_1. \tag{2}$$

Poisson's equation gives

$$\nabla \cdot \mathbf{E}_1 = -4\pi e n_1. \tag{3}$$

From equations (1)—(3), one derives

$$\frac{\partial^2 n_1}{\partial t^2} + \omega_p^2 n_1 = 0 \tag{4a}$$

or

$$n_1(t) = \tilde{n}_1 e^{\pm i\omega_p t} \tag{4b}$$

which describes the plasma oscillations, (or Langmuir waves). The restoring force for these oscillations is provided by the Coulomb interactions.

The first experiment to test (4b) was that of Looney and Brown [212]. They used an electron beam to excite plasma waves. The idea was that if the electrons in the beam were bunched so that they passed by any fixed point at a frequency f_p, they would generate an electric field at that frequency and excite plasma waves. Figure A.2.1 shows their experimental results for f^2 vs. discharge current (or density), the linear dependence shown by the points roughly agrees with (4b). Deviations from the straight line can be traced to the thermal effects.

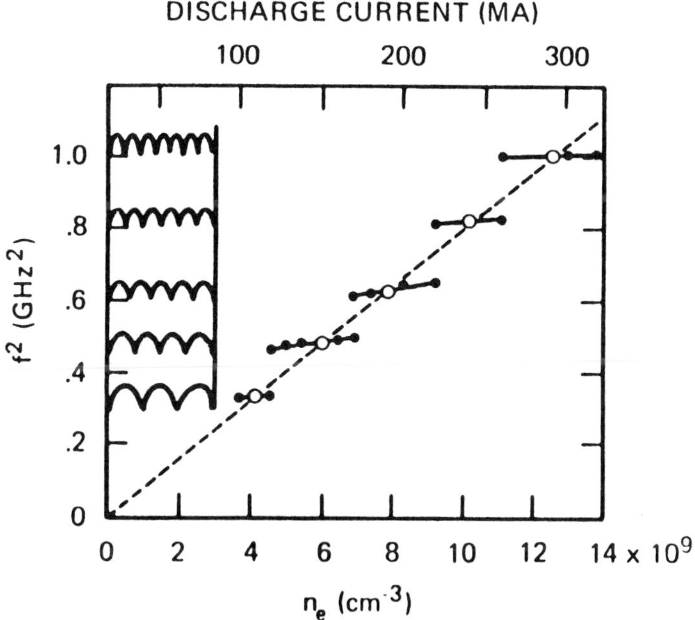

Figure A.2.1. Square of the observed frequency vs. plasma density, which is generally proportional to the discharge current. The inset shows the observed spatial distribution of oscillation intensity, indicating the existence of a different standing wave pattern for each of the groups of experimental points. (From D. H. Looney and S. C. Brown [212], by courtesy of The American Physical Society.)

Consider next the effect of the finite temperature of the plasma on the electron oscillations. Assume that the electrons form a charged fluid obeying the basic equations of hydrodynamics:

$$\frac{\partial n_1}{\partial t} + n_0 \nabla \cdot \mathbf{V}_1 = 0 \tag{5}$$

$$m_e n_0 \frac{\partial \mathbf{V}_1}{\partial t} = -\nabla p_1 - n_0 e \mathbf{E}_1 \tag{6}$$

along with Maxwell's equations:

$$\nabla \times \mathbf{B}_1 = -\frac{4\pi n_0 e}{c} \mathbf{V}_1 + \frac{1}{c}\frac{\partial \mathbf{E}_1}{\partial t} \tag{7}$$

$$\nabla \times \mathbf{E}_1 = -\frac{1}{c}\frac{\partial \mathbf{B}_1}{\partial t}. \tag{8}$$

The defect in the fluid model is the assumption as in equation (6) that the pressure is isotropic. This would be justified if $v_e \gg \omega_{p_e}$, where v_e is the collision frequency, but this is rarely true. To be rigorous, one really must adopt the kinetic approach, (which we will not do in this book).

One needs to use an equation of state:

$$\frac{\mathrm{d}}{\mathrm{d}t}\left(\frac{p_1}{\rho_1^{\gamma}}\right) = 0, \tag{9}$$

γ being the ratio of specific heats.

Using equation (5), equation (9) gives:

$$\frac{\partial p_1}{\partial t} + \gamma p_0 \nabla \cdot \mathbf{V}_1 = 0. \tag{10}$$

Let,

$$\mathbf{V}_1 = \frac{\partial \xi}{\partial t} \tag{11}$$

and all perturbations be of the form $\exp[i(\mathbf{k}\cdot\mathbf{x} - \omega t)]$; then one obtains from equations (5)–(8), (10) and (11):

$$\omega^2 \xi = \frac{\gamma p_0}{\rho_0} k^2 \xi + \frac{e}{m_e}\mathbf{E}_1 \tag{12}$$

$$c^2\mathbf{k} \times (\mathbf{k} \times \xi) = 4\pi n_0 e \omega^2 \xi - \omega^2 \xi. \tag{13}$$

For longitudinal oscillations, $(\mathbf{k} \times \mathbf{E}_1 = \mathbf{0})$, one obtains from (12) and (13):

$$\omega^2 = \omega_p^2 + k^2 V_T^2 \tag{14}$$

where,

$$V_T^2 = \frac{\gamma p_0}{\rho_0}.$$

For transverse oscillations, $(\mathbf{k} \cdot \mathbf{E}_1 = 0)$, one obtains from (12) and (13):

$$(\omega^2 - c^2k^2)(\omega^2 - k^2V_T^2) - \omega^2\omega_p^2 = 0. \tag{15}$$

If $\omega/k \gg V_T$, then (15) gives:

$$\omega^2 = w_p^2 + k^2c^2. \tag{16}$$

The dispersion relation (14) has been verified by a number of experiments. One such experiment was that of Barrett *et al.* [213] which used a cylindrical column of quiescent plasma produced in a Q-machine by thermal ionization of *Cs* atoms on hot tungsten plates. A strong magnetic field restricted electrons to move along the column. The waves were excited by a wire probe driven by an oscillator and were detected by a second, movable probe. The oscillator frequency was varied to generate a plot of the dispersion curve $(\omega/\omega_p)^2$ vs. ka, where a is the radius of the column (Figure A.2.2). The various curves correspond to different values of $\omega_p a/V_T$. Corresponding to $V_T = 0$, we have the curve labeled ∞, which corresponds to the dispersion relation $\omega = \omega_p$. Observe that there is good agreement between the experimental points and the theoretical curves. The decrease of ω at small values of ka represents the finite-geometry effect.

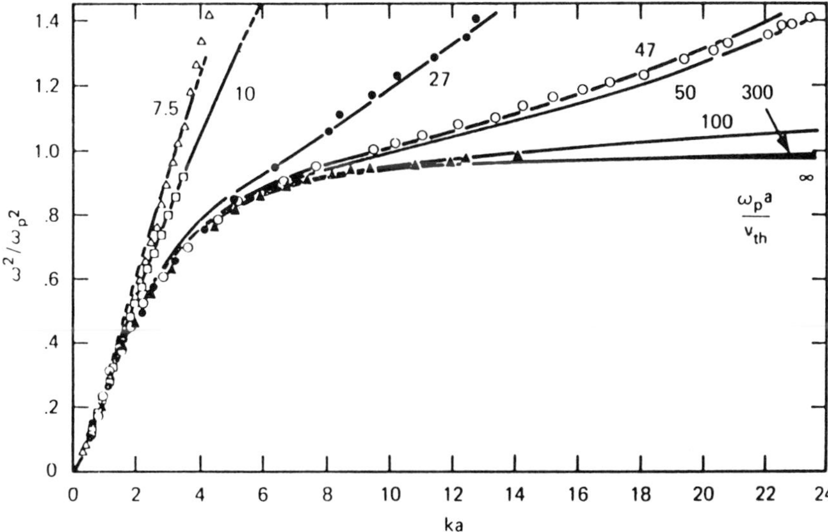

Figure A.2.2. Comparison of the measured and calculated dispersion curves for electron plasma waves in a cylinder of radius a. (From Barrett *et al.* [213], by courtesy of The Institute of Physics.)

In a tenuous plasma, the kinetic-theory approach (see for instance, Krall and Trivelpiece [119]) shows that the real part of the frequency is indeed given by that for the fluid model (14). But, the fluid model, which is a truncated set of moment equations obtained from the kinetic equations, completely misses the Landau damping of the plasma oscillations even in a collisionless plasma.

It should be noted that the longitudinal and transverse oscillations are strictly uncoupled in the case of a nonrelativistic plasma and in the absence of any external magnetic fields, or inhomogeneities in the plasma.

(ii) Ion-Acoustic Waves

In the absence of collisions, ordinary sound waves would not occur in a fluid. Ions can still transmit vibrations to each other because of their charge, however; and acoustic waves can occur through the intermediary of an electric field.

Electron-plasma oscillations in the foregoing were considered too rapid for the heavy ions to follow, so that the latter were taken to remain stationary. In constrast, the ion-acoustic waves are so slow that the electrons find them as being quasi-static and consequently remain distributed according to the Boltzmann distribution. We consider here longitudinal oscillations only.

The linearized equations for the ion-motions are:

$$\frac{\partial n_{i_1}}{\partial t} + n_0 \nabla \cdot \mathbf{V}_1 = 0 \tag{17}$$

$$m_i n_0 \frac{\partial \mathbf{V}_1}{\partial t} = -\nabla p_1 + n_0 e \mathbf{E}_1 \tag{18}$$

$$\nabla \cdot \mathbf{E}_1 = 4\pi e (n_{i_1} - n_{e_1}) \tag{19}$$

and for the electrons,

$$n_{e_1} = n_0 \left(\exp\left[\frac{e\phi_1}{KT_e} \right] - 1 \right) \tag{20}$$

where,

$$\mathbf{E}_1 = -\nabla \phi_1.$$

Assume a space-time dependence for the perturbations of the form $\exp[i(\mathbf{k} \cdot \mathbf{x} - \omega t)]$ so that equations (19) and (20) give:

$$\phi_1 = \frac{4\pi e n_{i_1}}{k^2 + 1/\lambda_D^2}, \quad \lambda_D^2 = \frac{KT_e}{4\pi n_0 e^2}. \tag{21}$$

Using (21), equations (17) and (18) give if $T_i \ll T_e$,

$$\frac{\omega^2}{k^2} = \frac{KT_e/m_i}{1 + k^2 \lambda_D^2}. \tag{22}$$

If, $k\lambda_D \ll 1$, (22) gives:

$$\frac{\omega}{k} \approx \sqrt{\frac{KT_e}{m_i}} = C_s. \tag{23}$$

If, $k\lambda_D \gg 1$, (22) gives:

$$\omega \approx \omega_{p_i} \tag{24}$$

the ion oscillations are then unshielded.

When $T_e \gg T_i$, the electrons are so mobile that they can immediately shield the ions. The electric field required to hold the electrons in place, against their own thermal motions, acts also on the ions, whose thermal motions are negligible. This provides the necessary restoring force for the ion-acoustic oscillations.

The Langmuir waves and ion-acoustic waves are schematically sketched in the ω, k-plane in Figure A.2.3.

Experimental verification of the existence of ion-acoustic waves was first accomplished by Wong *et al.* [214] in a quiescent plasma in a Q-machine. The phase velocities V_ϕ of the waves are plotted in Figure A.2.4 for the two masses and various plasma densities n_0. Observe the constancy of V_ϕ with ω and n_0, and that the two sets of points for K and C_s plasmas show the proper dependence on m_i.

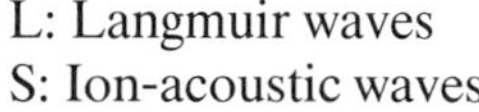

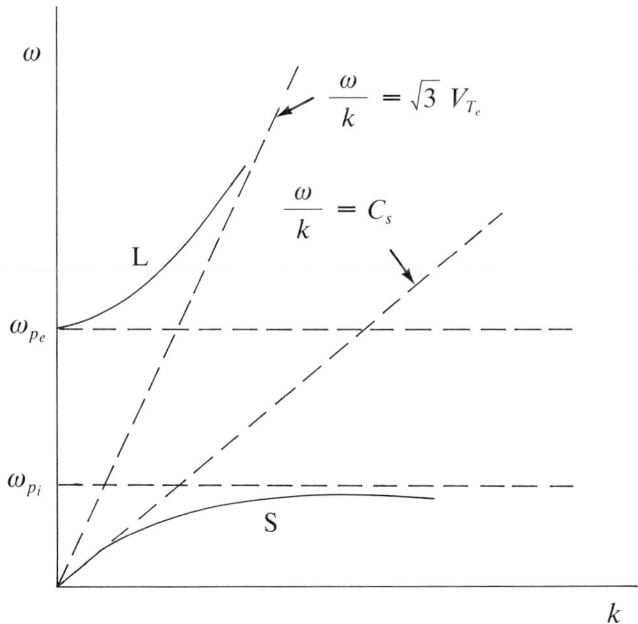

Figure A.2.3

(iii) Waves in a Magnetized Plasma

Since the behavior of a magnetized plasma becomes quite complex (for one thing, a magnetized plasma becomes anisotropic), in the interests of simplicity, we here neglect the thermal motions of the charged particles and consider the plasma as being cold.

The linearized equations for the perturbations about a static equilibrium are:

$$\frac{\partial \mathbf{V}_1}{\partial t} = \frac{e}{m}\left(\mathbf{E}_1 + \frac{1}{c}\,\mathbf{V}_1 \times \mathbf{B}_0\right) \qquad (25)$$

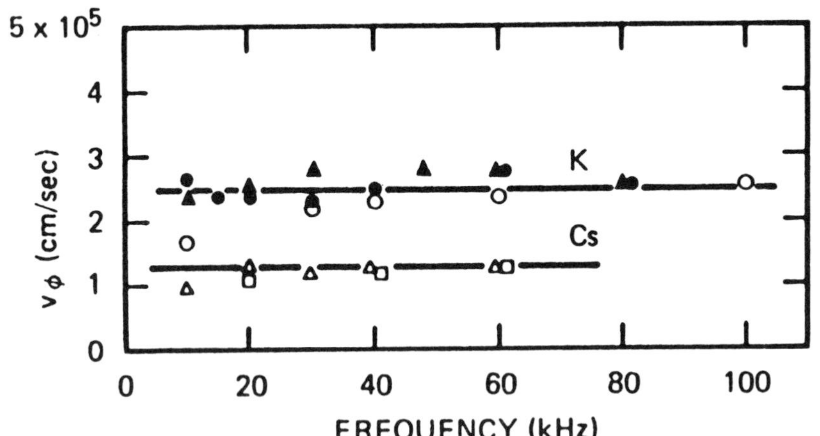

Figure A.2.4. Measured phase velocity of ion waves in potassium and cesium plasmas as a function of frequency. The different sets of points correspond to different plasma densities. (From Wong *et al.* [214], by courtesy of The American Physical Society.)

$$\frac{\partial n_1}{\partial t} + n_0 \nabla \cdot \mathbf{V}_1 = 0 \tag{26}$$

$$\nabla \times \mathbf{B}_1 = \frac{1}{c}\frac{\partial \mathbf{E}_1}{\partial t} + \frac{4\pi}{c}\sum_s e n_0 \mathbf{V}_1 \tag{27}$$

$$\nabla \times \mathbf{E}_1 = -\frac{1}{c}\frac{\partial \mathbf{B}_1}{\partial t} \tag{28}$$

where s refers to the electron and ion species, and the subscript 1 refers to the perturbations.

Assume that the perturbations have a space-time dependence of the form $\exp[i(\mathbf{k} \cdot \mathbf{x} - \omega t)]$. Then, equations (25) and (26) give:

$$-i\omega \mathbf{V}_1 = \frac{e}{m}\mathbf{E}_1 - \mathbf{\Omega} \times \mathbf{V}_1 \tag{29}$$

$$n_1 = \frac{n_0}{\omega}\mathbf{k} \cdot \mathbf{V}_1 \tag{30}$$

where

$$\mathbf{\Omega} = \frac{e\mathbf{B}_0}{mc}.$$

From (29), one obtains:

$$-i\omega \mathbf{\Omega} \cdot \mathbf{V}_1 = \frac{e}{m}\mathbf{\Omega} \cdot \mathbf{E}_1 \tag{31}$$

$$-i\omega \mathbf{\Omega} \times \mathbf{V}_1 = \frac{e}{m} \mathbf{\Omega} \times \mathbf{E}_1 - \mathbf{\Omega}(\mathbf{\Omega} \cdot \mathbf{V}_1) + \Omega^2 \mathbf{V}_1. \tag{32}$$

From (31) and (32), one obtains:

$$\mathbf{V}_1 = \frac{e/m}{\omega^2 - \Omega^2} \left(i\omega \mathbf{E}_1 + \mathbf{\Omega} \times \mathbf{E}_1 - \frac{i\mathbf{\Omega}}{\omega} \mathbf{\Omega} \cdot \mathbf{E}_1 \right). \tag{33}$$

From equations (27) and (28), one obtains:

$$(\omega^2 - k^2c^2)\mathbf{E}_1 + c^2\mathbf{k}(\mathbf{k} \cdot \mathbf{E}_1) + 4\pi i\omega \sum_s en_0 \mathbf{V}_1 = 0. \tag{34}$$

Using (33), (34) gives:

$$\mathscr{R} \cdot \mathbf{E}_1 = \mathbf{0} \tag{35}$$

where

$$\mathscr{R} = \begin{pmatrix} \mathscr{R}_{11} & \mathscr{R}_{12} & \mathscr{R}_{13} \\ \mathscr{R}_{21} & \mathscr{R}_{22} & \mathscr{R}_{23} \\ \mathscr{R}_{31} & \mathscr{R}_{32} & \mathscr{R}_{33} \end{pmatrix}$$

$$\mathscr{R}_{11} = \omega^2 - c^2k^2 - \sum_s \omega_p^2 \left(1 - \frac{\Omega^2 \cos^2 \theta}{\Omega^2 - \omega^2} \right)$$

$$\mathscr{R}_{12} = -\mathscr{R}_{21} = \sum_s i\omega_p^2 \frac{\omega\Omega \cos \theta}{\Omega^2 - \omega^2}$$

$$\mathscr{R}_{13} = -\mathscr{R}_{31} = -\sum_s \omega_p^2 \frac{\Omega^2 \sin \theta \cos \theta}{\Omega^2 - \omega^2}$$

$$\mathscr{R}_{22} = \omega^2 - c^2k^2 + \sum_s \frac{\omega_p^2\omega^2}{\Omega^2 - \omega^2}$$

$$\mathscr{R}_{23} = -\mathscr{R}_{32} = i \sum_s \omega_p^2 \frac{\omega\Omega \sin \theta}{\Omega^2 - \omega^2}$$

$$\mathscr{R}_{33} = \omega^2 - \sum_s \omega_p^2 \left(1 - \frac{\Omega^2 \sin^2 \theta}{\Omega^2 - \omega^2} \right).$$

Here **k** is taken to be along the z-axis, and $\mathbf{B}_0$ lies in the xz-plane at an angle θ with the z-axis. In order that equation (35) has a nontrivial solution, one requires:

$$\|\mathscr{R}\| = 0 \tag{36}$$

which is the dispersion relation. For an arbitrary direction of **k** with respect to

$\mathbf{B}_0$, it is not easy to straighten out (36). The essential features of the problem can be brought forth by considering the special cases of propagation along the magnetic field and transverse to the magnetic field.

(a) Propagation Along the Magnetic Field

One obtains from (35) for $\theta = 0$,

$$\begin{bmatrix} \omega^2 - c^2k^2 + \sum_s \frac{\omega_p^2\omega^2}{\Omega^2 - \omega^2} & i\sum_s \omega_p^2 \frac{\omega\Omega}{\Omega^2 - \omega^2} & 0 \\ -i\sum_s \omega_p^2 \frac{\omega\Omega}{\Omega^2 - \omega^2} & \omega^2 - c^2k^2 + \sum_s \frac{\omega_p^2\omega^2}{\Omega^2 - \omega^2} & 0 \\ 0 & 0 & \omega^2 - \sum_s \omega_p^2 \end{bmatrix} \begin{bmatrix} E_{1_x} \\ E_{1_y} \\ E_{1_z} \end{bmatrix} = 0. \tag{37}$$

The longitudinal oscillations are given by

$$\left(\omega^2 - \sum_s \omega_p^2\right) E_{1_z} = 0 \tag{38}$$

from which,

$$\omega^2 \approx \omega_{p_e}^2. \tag{39}$$

The transverse oscillations are polarized, and are given by

$$\left(\omega^2 - c^2k^2 + \sum_s \omega_p^2 \frac{\omega}{\Omega \pm \omega}\right)(E_{1_x} \mp iE_{1_y}) = 0 \tag{40}$$

where the $\pm$ signs correspond, respectively, to the right- or left-handed circularly polarized waves which exhibit the ion- or electron-cyclotron resonance $\omega = \pm\Omega$. Near the resonances, the group velocity $d\omega/dk \Rightarrow 0$ (since $k \Rightarrow \infty$) so that the waves remain localized to a particular region of physical space for a long time. In general, various damping mechanisms, not present in 'cold' plasma model, are particularly strong near resonances.

Consider the right-handed circularly-polarized waves; one has from (40):

$$\omega^2 - c^2k^2 - \omega_{p_e}^2 \frac{\omega}{\omega + \Omega_e} - \omega_{p_i}^2 \frac{\omega}{\omega + \Omega_i} = 0. \tag{41}$$

If $\omega \gg \Omega_e, \Omega_i$, (41) gives:

$$\omega^2 \approx c^2k^2 + \omega_{p_e}^2\left(1 - \frac{\Omega_e}{\sqrt{\omega_{p_e}^2 + c^2k^2}}\right). \tag{42}$$

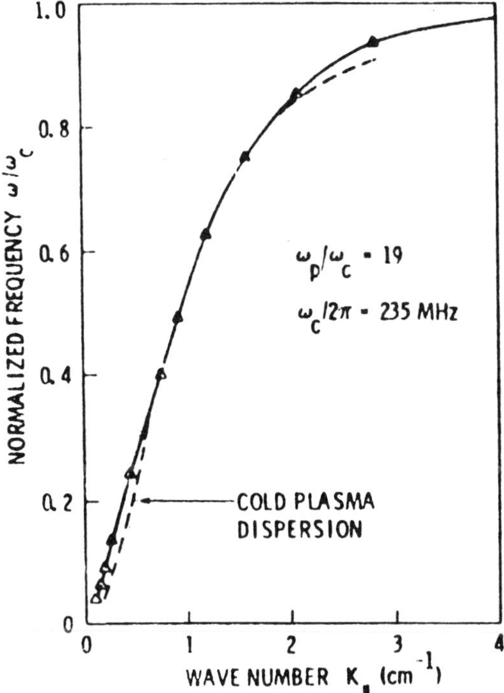

Figure A.2.5. Whistler wave dispersion relation. The solid line with data points are measured, the dashed line is the theoretical dispersion from Eq. (41). (From Stenzel [215], by courtesy of The American Institute of Physics.)

If $\omega \approx -\Omega_e \gg \Omega_i$, (41) gives:

$$\omega \approx -\Omega_e \left[1 + \frac{\omega_{p_e}^2}{\Omega_e^2 - (c^2 k^2 + \omega_{p_e}^2)} \right]. \tag{43}$$

If $\omega \ll \Omega_e$, (41) gives:

$$k \approx \frac{\omega_{p_e}}{c} \sqrt{\frac{\omega}{|\Omega_e|}} \tag{44}$$

which indicates the high-frequency components moving faster — these are the whistler waves, which are audio-frequency electromagnetic disturbances initiated in the atmosphere by lightning flashes. The signal received is a descending audio tone.

Stenzel [215] made detailed measurements on antenna-launched whistler waves in a large, collisionless, quiescent plasma column in the parameter regime $0.05 < \omega/\Omega_e < 1$, $\Omega_e/2\pi \sim 250$ MHz, $\omega_{p_e}/\Omega_e \gtrsim 10$. Both the antenna characteristics and the wave propagation effects were observed by mapping the spatial field distribution. By varying the frequency and measuring the wavelength of the small-amplitude wave at large distances from the exciter antenna the dispersion diagram ω vs. k (Figure A.2.5) was obtained. It is seen that over a wide range of frequencies ($0.3 < \omega/\Omega_e < 0.8$) the observed dispersion relation is well approximated by the theoretical expression (41) (without the ion term). The deviations at high frequencies ($0.9 < \omega/\Omega_e < 1.0$) reflect the increasing discrepancy between antenna-excited diverging waves and plane waves (considered in (41)) as the antenna size becomes small compared with the wavelength.

If $\omega \ll \Omega_i$, (41) gives

$$\omega^2 \approx \frac{c^2k^2}{1 + c^2/V_A^2} \tag{45}$$

where

$$V_A^2 = \frac{B_0^2}{4\pi n_0 m_i}.$$

If $V_A/c \ll 1$, (45) gives the Alfvén waves:

$$\omega/k \approx V_A \tag{46}$$

Note that at low frequencies, the underlying motion is that of the ions; the electrons serving only to re-establish electrical neutrality whenever it is violated.

Alfvén waves have been generated and detected in a number of experiments. One such experiment was that of Wilcox *et al.* [216] which used a hydrogen plasma created in a 'slow pinch' discharge between two electrodes aligned along a magnetic field. Figure A.2.6 shows measurements of phase velocity vs. magnetic field, demonstrating the linear dependence predicted by (46).

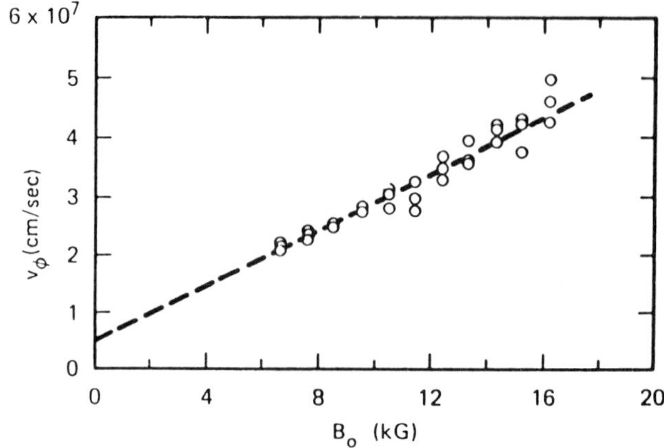

Figure A.2.6. Measured phase velocity of Alfvén waves vs. magnetic field. (From Wilcox *et al.* [216], by courtesy of The American Institute of Physics.)

If the amplitudes of both circularly-polarized waves are equal initially, one may superpose them to produce a plane wave. Because the two polarizations propagate at different speeds, the plane of polarization rotates as the wave propagates along the magnetic field — this is called Faraday rotation.

The two circularly-polarized waves are shown in Figure A.2.7.

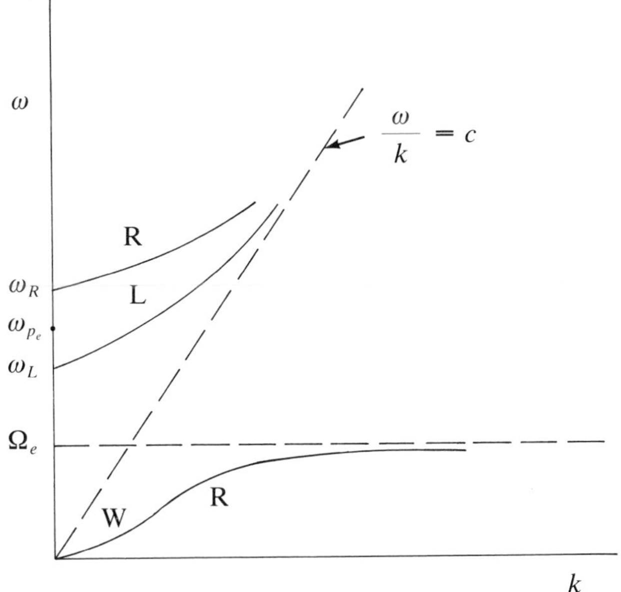

Figure A.2.7.

(*b*) *Propagation Transverse to the Magnetic Field*

If the waves propagate at an angle $\theta \approx \pi/2$ with respect to the magnetic field, one may take $\theta = \pi/2$ for the ion motions. But it makes a great deal of difference whether $\pi/2 - \theta$ is zero, or small but finite, insofar as the electron motions are concerned. The electrons have such small Larmor radii that they cannot move in the x-direction to preserve the charge neutrality. If θ is not exactly $\pi/2$, however, the electrons can move along B_0 to bring about charge neutrality. (The ions cannot do this effectively because their inertia prevents them from moving long distances in a wave period.)

One obtains from (35) for $\theta = \pi/2$,

$$\begin{bmatrix} \omega^2 - c^2k^2 - \sum_s \omega_p^2 & 0 & 0 \\ 0 & \omega^2 - c^2k^2 + \sum_s \frac{\omega_p^2\omega^2}{\Omega^2 - \omega^2} & i\sum_s \omega_p^2 \frac{\omega\Omega}{\Omega^2 - \omega^2} \\ 0 & -i\sum_s \omega_p^2 \frac{\omega\Omega}{\Omega^2 - \omega^2} & \omega^2 + \sum_s \frac{\omega_p^2\omega^2}{\Omega^2 - \omega^2} \end{bmatrix} \begin{bmatrix} E_{1_x} \\ E_{1_y} \\ E_{1_z} \end{bmatrix} = 0. \quad (47)$$

From (47), one has the usual transverse oscillations:

$$\omega^2 = c^2k^2 + \sum_s \omega_p^2 \approx c^2k^2 + \omega_{p_e}^2 \quad (48)$$

which are plane-polarized 'ordinary' waves since associated with them is $\mathbf{E}_1$ parallel to $\mathbf{B}_0$.

For the other two modes of oscillations, which are elliptically polarized 'extraordinary' waves since now $\mathbf{E}_1 \cdot \mathbf{B}_0 = 0$, one has

$$c^2k^2 = \omega^2 + \sum_s \frac{\omega_p^2\omega^2}{\Omega^2 - \omega^2} - \frac{\left(\sum_s \frac{\omega_p^2\omega^2}{\Omega^2 - \omega^2}\right)^2}{1 + \sum_s \frac{\omega_p^2}{\Omega^2 - \omega^2}} \quad (49)$$

or

$$\frac{c^2k^2}{\omega^2} = \frac{\left(1 - \frac{\omega_R^2}{\omega^2}\right)\left(1 - \frac{\omega_L^2}{\omega^2}\right)}{\left(1 - \frac{\omega_{\mathrm{LH}}^2}{\omega^2}\right)\left(1 - \frac{\omega_{\mathrm{UH}}^2}{\omega^2}\right)} \quad (50)$$

where,

$$\omega_{R,L} = \frac{\Omega_e}{2}\left[\left(1 + \frac{4\omega_{p_e}^2}{\Omega_e^2}\right)^{1/2} \pm 1\right]$$

$$\omega_{\mathrm{LH}}^2 = \frac{\omega_{p_i}^2}{1 + \omega_{p_e}^2/\Omega_e^2} \approx -\Omega_e\Omega_i$$

$$\omega_{\mathrm{UH}}^2 = \omega_{p_e}^2 + \Omega_e^2.$$

Equation (50) shows resonances at $\omega \approx \omega_{\mathrm{LH}}$, and ω_{UH}.

For $\omega \ll ck$, (50) gives:

$$\begin{aligned} \omega &\approx \omega_{\mathrm{LH}} \quad \text{if} \quad \Omega_i < \omega < \Omega_e, \\ \omega &\approx \omega_{\mathrm{UH}} \quad \text{if} \quad \omega \gg \Omega_i, \quad \omega_{p_i}. \end{aligned} \tag{51}$$

ω_{LH} and ω_{UH} are called, respectively, the lower-hybrid and upper-hybrid frequencies. It can be verified that for $\omega \approx \omega_{\mathrm{UH}}$, the wave is almost entirely longitudinal.

If $\omega \ll \Omega_i$, (49) gives:

$$\frac{\omega^2}{k^2} \approx \frac{V_A^2}{1 + V_A^2/c^2} \tag{52}$$

which is a magnetosonic wave.

The waves with $\mathbf{k} \cdot \mathbf{B}_0 = 0$ are shown in Figure A.2.8.

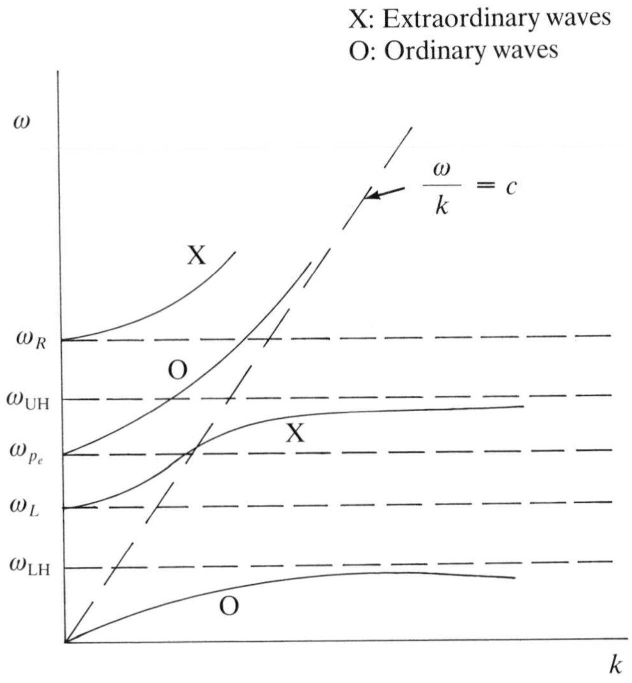

Figure A.2.8.

The existence of the upper hybrid frequency has been verified experimentally by microwave transmission across a magnetic field. As the plasma density changes, the transmission through the plasma shows a drop at the layer where ω_{UH} is equal to the applied frequency. This is due to the excitation of the upper-hybrid oscillations and absorption of energy from the beam. Note that

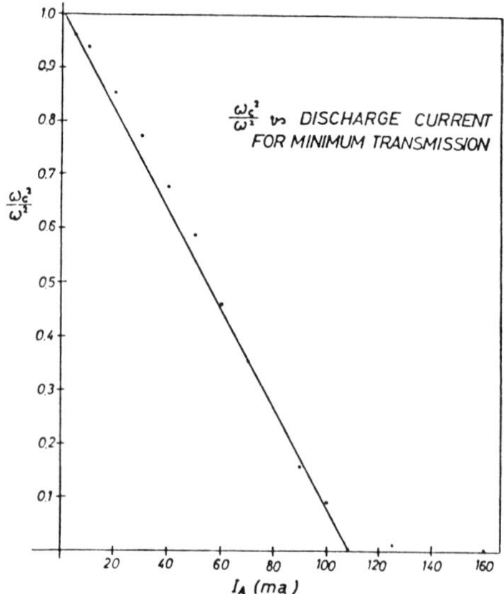

Figure A.2.9. Results of an experiment to detect the existence of the upper hybrid frequency by mapping the conditions for maximum absorption (minimum transmission) of microwave energy sent across a magnetic field. The field at which this occurs (expressed as ω_c^2/ω^2) is plotted against discharge current (proportional to plasma density). (From R. S. Harp [217], by courtesy of The International Atomic Energy Agency.)

the linear relationship between Ω_e^2/ω^2 and the density from (50):

$$\frac{\Omega_e^2}{\omega^2} = 1 - \frac{\omega_{p_e}^2}{\omega^2} = 1 - \frac{ne^2}{m\omega^2} .$$

This linear relation was verified by the experimental results of Harp [217] shown in Figure A.2.9 where Ω_e^2/ω^2 is plotted against the discharge current (or the density).

References

[1] J. M. Dawson: *Rev.* **113**, 383 (1959).
[2] R. C. Davidson and P. C. Schram: *Nucl. Fusion* **8**, 183 (1968).
[3] J. Albritton and G. Rowlands: *Nucl. Fusion* **15**, 1199 (1975).
[4] H. Washimi and T. Taniuti: *Phys. Rev. Lett.* **17**, 996 (1966).
[5] R. Z. Sagdeev: *Rev. Plasma Phys.* **4**, 23 (1966).
[6] N. Zabusky and M. D. Kruskal: *Phys. Rev. Lett.* **151**, 247 (1965).
[7] C. S. Gardner, J. M. Greene, M. D. Kruskal, and R. M. Muira: *Phys. Rev. Lett.* **19**, 1095 (1967).
[8] H. Ikezi, R. J. Taylor, and R. D. Baker: *Phys. Rev. Lett.* **25**, 11 (1970).
[9] H. Ikezi: *Phys. Fluids* **16**, 1668 (1973).
[10] S. G. Alikhanov, V. G. Belan, and R. Z. Sagdeev: *Zh. Eksp. Theor. Fiz. Pis'ma Red.* **7**, 405 (1968).
[11] R. J. Taylor, R. D. Baker and H. Ikezi: *Phys. Rev. Lett.* **24**, 206 (1970).
[12] V. P. Silin: *Sov. Phys. JETP* **21**, 1127 (1965).
[13] D. F. DuBois and M. V. Goldman: *Phys. Rev. Lett.* **14**, 544 (1965).
[14] A. Sjolund and Stenflo: *Physica* **35**, 499 (1967).
[15] B. K. Shivamoggi: *J. Tech. Phys.* **17**, 281 (1975).
[16] C. N. Lashmore-Davies: *Plasma Phys.* **17**, 281 (1975).
[17] A. Sjolund and L. Stenflo: *Z. Physik* **204**, 211 (1967).
[18] C. S. Chen and G. J. Lewak: *J. Plasma Phys.* **4**, 357 (1970).
[19] R. Prasad: *J. Plasma Phys.* **5**, 291 (1971).
[20] K. F. Lee: *J. Plasma Phys.* **11**, 99 (1974).
[21] B. K. Shivamoggi: *Fizika* **16**, 167 (1984).
[22] L. Stenflo: *J. Plasma Phys.* **5**, 413 (1971).
[23] L. Stenflo: *J. Plasma Phys.* **7**, 107 (1972).
[24] S. Krishan, D. Rankin, and A. A. Selim: *J. Plasma Phys.* **9**, 23 (1973).
[25] P. Munoz and S. Dagach: *J. Plasma Phys.* **17**, 51 (1977).
[26] K. J. Harker and F. W. Crawford: *J. Appl. Phys.* **40**, 3247 (1969).
[27] K. P. Das: *Phys. Fluids* **14**, 124 (1971).
[28] M. Porkolab and R. P. H. Chang: *Phys. Fluids* **13**, 2766 (1970).
[29] K. Phelps, N. Rynn, and G. Van Hoven: *Phys. Rev. Lett.* **26**, 688 (1971).
[30] R. N. Franklin, S. M. Hamberger, H. Ikezi, G. Lampis, and G. J. Smith: *Phys. Rev. Lett.* **27**, 1119 (1971).
[31] D. Montgomery and I. Alexeff: *Phys. Fluids* **9**, 1362 (1966).
[32] K. Nishikawa: *J. Phys. Soc. Japan* **24**, 916 (1968).
[33] M. Bornatici: *J. Plasma Phys.* **14**, 105 (1975).
[34] B. K. Shivamoggi: *Austral. J. Phys.* **35**, 409 (1982).

[35] B. H. Ripin, J. M. McMahon, R. A. McLean, W. M. Manheimer, and J. A. Stamper: *Phys. Rev. Lett.* **33**, 634 (1974).
[36] A. A. Offenberger, M. R. Cervenan, A. M. Yam, and A. W. Pasternak: *J. Appl. Phys.* **47**, 1451 (1976).
[37] A. A. Offenberger, R. Fedosejevs, W. Tighe and W. Rozmus: *Phys. Rev. Lett.* **49**, 371 (1982).
[38] J. J. Turechek and F. F. Chen: *Phys. Rev. Lett.* **36**, 720 (1976).
[39] B. Grek, H. Pepin, and F. Rheault: *Phys. Rev. Lett.* **38**, 898 (1977).
[40] R. G. Watt, R. D. Brooks, and Z. A. Pietrzyk: *Phys. Rev. Lett.* **41**, 170 (1978).
[41] A. Ng, D. Salzmann, and A. A. Offenberger: *Phys. Rev. Lett.* **42**, 307 (1979).
[42] E. J. Valeo, C. Oberman, and F. W. Perkins: *Phys. Rev. Lett.* **28**, 218 (1972).
[43] D. F. DuBois and M. V. Goldman: *Phys. Rev. Lett.* **28**, 218 (1972).
[44] V. E. Zakharov: *Sov. Phys. JETP* **26**, 994 (1968).
[45] F. E. Zakharov: *Sov. Phys. JETP* **35**, 908 (1972).
[46] M. V. Goldman and D. R. Nicholson: *Phys. Rev. Lett.* **41**, 406 (1978).
[47] N. R. Pereira, R. N. Sudan, and J. Denavit: *Phys. Fluids* **20**, 936 (1977).
[48] D. R. Nicholson, M. V. Goldman, P. Hoyng, and J. C. Weatherall: *Astrophys. J.* **22**, 605 (1978).
[49] B. Hafizi, J. C. Weatherall, M. V. Goldman, and D. R. Nicholson: *Phys. Fluids* **25**, 392 (1982).
[50] G. Schmidt: *Phys. Rev. Lett.* **34**, 724 (1975).
[51] V. I. Karpman: *Plasma Phys.* **13**, 477 (1971).
[52] V. I. Karpman: *Phys. Scripta* **11**, 263 (1975).
[53] L. I. Rudakov: *Sov. Phys. Dokl.* **17**, 1166 (1973).
[54] V. E. Zakharov and A. B. Shabat: *Sov. Phys. JETP* **34**, 62 (1972).
[55] G. J. Morales, Y. C. Lee, and R. B. White: *Phys. Rev. Lett.* **32**, 457 (1974).
[56] A. Thyagaraja: *Phys. Fluids* **22**, 2093 (1979).
[57] A. Thyagaraja: *Phys. Fluids* **24**, 1973 (1981).
[58] J. Gibbons, S. G. Thornhill, M. J. Wardrop, and D. ter Haar: *J. Plasma Phys.* **17**, 453 (1977).
[59] V. G. Makhankov: *Phys. Lett.* **50A**, 42 (1974).
[60] K. Nishikawa, H. Hojo, K. Mima, and H. Ikezi: *Phys. Rev. Lett.* **33**, 149 (1975).
[61] N. N. Rao and R. K. Varma: *J. Plasma Phys.* **27**, 195 (1982).
[62] B. K. Shivamoggi: *Can. J. Phys.* **61**, 1205 (1983).
[63] B. K. Shivamoggi: To be published (1988).
[64] E. Garmire, R. Y. Chiao, and C. H. Townes: *Phys. Rev. Lett.* **16**, 347 (1966).
[65] A. J. Campillo, S. L. Shapiro, and B. R. Suydam: *Appl. Phys. Lett.* **23**, 628 (1973).
[66] H. Ikezi, K. Nishikawa, and K. Mima: *J. Phys. Soc. Japan* **37**, 766 (1974).
[67] H. Ikezi, R. P. H. Chang, and R. A. Stern: *Phys. Rev. Lett.* **36**, 1047 (1976).
[68] H. C. Kim, R. L. Stenzel, and A. Y. Wong: *Phys. Rev. Lett.* **33**, 886 (1974).
[69] P. Y. Cheung, A. Y. Wong, C. B. Darrow, and S. J. Qian: *Phys. Rev. Lett.* **48**, 1348 (1982).
[70] A. Y. Wong and B. H. Quon: *Phys. Rev. Lett.* **34**, 1499 (1975).
[71] P. Leung, M. Q. Tran, and A. Y. Wong: *Plasma Phys.* **24**, 567 (1982).
[72] A. S. Kingsep, L. I. Rudakov, and R. N. Sudan: *Phys. Rev. Lett.* **31**, 1482 (1973).
[73] A. A. Galeev, R. Z. Sagdeev, Y. S. Sigov, V. D. Shapiro, and V. I. Shevchenko: *Sov. J. Plasma Phys.* **1**, 5 (1975).
[74] A. A. Galeev, R. Z. Sagdeev, V. D. Shapiro, and V. I. Shevchenko: *Sov. Phys. JETP* **46**, 711 (1978).
[75] A. Y. Wong and P. Y. Cheung: *Phys. Fluids* **28**, 1538 (1985).
[76] L. I. Rudakov and V. N. Tsytovich: *Phys. Rept.* **40**, 1 (1978).
[77] S. G. Thornhill and D. ter Haar: *Phys. Rept.* **43**, 43 (1978).
[78] K. Shimizu and Y. H. Ichikawa: *J. Phys. Soc. Japan* **33**, 789 (1972).
[79] T. Kakutani and N. Sugimoto: *Phys. Fluids* **17**, 1617 (1974).
[80] V. S. Chan and S. R. Seshadri: *Phys. Fluids* **20**, 1294 (1975).
[81] E. Infeld and G. Rowlands: *J. Plasma Phys.* **25**, 81 (1981).

[82] B. K. Shivamoggi: *Can. J. Phys.* **63**, 435 (1985).
[83] H. Ikezi, K. Schwarzenegger, A. L. Simmons, Y. Ohsawa, and T. Kamimura: *Phys. Fluids* **21**, 239 (1978).
[84] A. N. Kaufmann and L. Stenflo: *Phys. Scripta* **11**, 269 (1975).
[85] M. Porkolab and M. V. Goldman: *Phys. Fluids* **19**, 872 (1976).
[86] B. K. Shivamoggi: *Phys. Scripta* **26**, 401 (1982).
[87] T. Cho and S. Tanaka: *Phys. Rev. Lett.* **45**, 1403 (1980).
[88] W. Gekelman and R. L. Stenzel: *Phys. Rev. Lett.* **35**, 1708 (1975).
[89] G. J. Morales and Y. C. Lee: *Phys. Rev. Lett.* **35**, 930 (1975).
[90] C. K. W. Tam: *Phys. Fluids* **12**, 1028 (1969).
[91] T. Taniuti and H. Washimi: *Phys. Rev. Lett.* **32**, 454 (1969).
[92] A. Hasegawa: *Phys. Rev. A* **1**, 1746 (1970).
[93] A. Hasegawa: *Phys. Fluids* **15**, 870 (1972).
[94] H. Washimi and V. I. Karpman: *Sov. Phys. JETP* **44**, 528 (1976).
[95] V. I. Karpman and H. Washimi: *J. Plasma Phys.* **18**, 173 (1977).
[96] A. Rogister: *Phys. Fluids* **14**, 2733 (1971).
[97] E. Mjolhus: *J. Plasma Phys.* **16**, 321 (1976).
[98] E. Mjolhus: *J. Plasma Phys.* **19**, 437 (1978).
[99] S. R. Spangler and J. P. Sheerin: *J. Plasma Phys.* **27**, 193 (1982).
[100] C. N. Lashmore-Davies and L. Stenflo: *Plasma Phys.* **21**, 735 (1979).
[101] A. I. Akhiezer and R. V. Polovin: *Sov. Phys. JETP* **3**, 696 (1956).
[102] W. Lunow: *Plasma Phys.* **10**, 879 (1968).
[103] P. K. Kaw and J. M. Dawson: *Phys. Fluids* **13**, 472 (1970).
[104] C. Max and F. W. Perkins: *Phys. Rev. Lett.* **27**, 1342 (1971).
[105] A. C. L. Chian and P. C. Clemmow: *J. Plasma Phys.* **14**, 505 (1975).
[106] A. DeCoster: *Phys. Rept.* **47**, 285 (1978).
[107] P. K. Kaw, A. Sen, and E. J. Valeo: *Physica D* **9**, 96 (1983).
[108] F. W. Sluijter, D. Montgomery: *Phys. Fluids* **8**, 551 (1965).
[109] J. A. Stamper, K. Papadopoulos, R. N. Sudan, S. O. Dean, E. A. McLean, and J. M. Dawson: *Phys. Rev. Lett.* **26**, 1012 (1971).
[110] J. A. Stamper and B. H. Ripin: *Phys. Rev. Lett.* **34**, 138 (1975).
[111] J. A. Stamper, E. A. McLean, and B. H. Ripin: *Phys. Rev. Lett.* **40**, 1177 (1978).
[112] D. J. Diverglio, A. Y. Wong, H. C. Kim, and Y. C. Lee: *Phys. Rev. Lett.* **38**, 541 (1977).
[113] Y. M. Aliev and S. V. Kuznetsov: *Sov. J. Plasma Phys.* **6**, 205 (1980).
[114] V. I. Berzhiani, V. S. Paverman, and D. D. Tskhakaya: *Sov. J. Plasma Phys.* **6**, 445 (1980).
[115] B. K. Shivamoggi: Unpublished (1987).
[116] C. Max, J. Arons, and A. B. Langdon: *Phys. Rev. Lett.* **33**, 209 (1974).
[117] B. K. Shivamoggi: *Phys. Rev. A.* **31**, 1728 (1985).
[118] T. P. Coffey: *Phys. Fluids* **14**, 1402 (1971).
[119] N. Krall and A. Trivelpiece: *Principles of Plasma Physics*, McGraw-Hill, New York (1973).
[120] A. C. Scott, F. Y. F. Chu, and D. W. McLaughlin: *Proc. IEEE* **61**, 1443 (1973).
[121] E. Ott and R. N. Sudan: *Phys. Fluids* **13**, 1432 (1970).
[122] E. Ott: *Phys. Fluids* **14**, 748 (1971).
[123] K. Abe and T. Abe: *Phys. Fluids* **22**, 1644 (1979).
[124] K. Abe and N. Satofuka: *Phys. Fluids* **24**, 1045 (1981).
[125] V. E. Zakharov and E. M. Kuznetsov: *Soviet Phys. JETP* **39**, 285 (1974).
[126] E. W. Laedke and K. H. Spatschek: *Phys. Fluids* **25**, 985 (1982).
[127] E. W. Laedke and K. H. Spatschek: *J. Plasma Phys.* **28**, 469 (1982).
[128] E. Infeld: *J. Plasma Phys.* **33**, 171 (1985).
[129] E. Infeld and P. Frycz: *J. Plasma Phys.* **37**, 97 (1987).
[130] B. K. Shivamoggi: *J. Plasma Phys.* (1988).
[131] R. Peierls: *Quantum Theory of Solids*, Oxford Univ. Press (1955).
[132] J. Weiland and H. Wilhelmsson: *Coherent Nonlinear Interactions of Waves in Plasmas*, Pergamon Press, London (1977).
[133] J. M. Manley and H. E. Rowe: *Proc. IRE* **47**, 2115 (1959).

[134] R. Sugihara: *Phys. Fluids* **11**, 178 (1968).
[135] J. P. Dougherty: *J. Plasma Phys.* **4**, 761 (1970).
[136] J. Galloway and H. Kim: *J. Plasma Phys.* **6**, 53 (1971).
[137] S. Krishan and J. Fukai: *Phys. Fluids* **14**, 1158 (1971).
[138] B. Coppi, M. N. Rosenbluth, and R. N. Sudan: *Ann. Phys.* (*NY*) **55**, 207 (1969).
[139] V. M. Dikasov, L. I. Rudakov, and D. D. Ryutov: *Sov. Phys. JETP* **21**, 608 (1966).
[140] R. Sugaya, M. Sugawa, and H. Nomoto: *Phys. Rev. Lett.* **39**, 27 (1977).
[141] F. F. Chen: *Introduction to Plasma Physics and Controlled Fusion Volume 1: Plasma Physics*, Plenum, New York (1984).
[142] B. Bezzerides, D. DuBois, D. W. Forslund, and E. L. Lindman: *Phys. Rev. A* **16**, 1678 (1977).
[143] S. Johnston, A. N. Kaufmann, and G. L. Johnston: *J. Plasma Phys.* **20**, 365 (1978).
[144] S. A. Belkov and V. N. Tsytovich: *Sov. Phys. JETP* **49**, 656 (1979).
[145] M. Kono, M. M. Skoric, and D. ter Haar: *Phys. Rev. Lett.* **45**, 1629 (1980).
[146] M. Kono, M. M. Skoric, and D. ter Haar: *J. Plasma Phys.* **26**, 123 (1981).
[147] M. C. Festeau-Barrioz and E. S. Weibel: *Phys. Fluids* **23**, 2045 (1980).
[148] D. D. Tskhakaya: *J. Plasma Phys.* **25**, 233 (1981).
[149] W. Washimi and V. I. Karpman: *Soviet Phys. JETP* **44**, 528 (1976).
[150] V. I. Karpman and A. G. Shagalov: *J. Plasma Phys.* **27**, 215 (1982).
[151] G. Statham and D. ter Haar: *Plasma Phys.* **25**, 681 (1983).
[152] B. M. Lamb, G. Dimonte, and G. J. Morales: *Phys. Fluids* **27**, 1401 (1984).
[153] H. P. Freund and K. Papadopoulos: *Phys. Fluids* **23**, 139 (1980).
[154] D. W. Forslund, J. M. Kindel, and E. L. Lindman: *Phys. Rev. Lett.* **29**, 249 (1972).
[155] K. F. Lee: *Phys. Fluids* **17**, 1343 (1974).
[156] V. K. Tripathi and C. S. Liu: *Phys. Fluids* **22**, 1761 (1979).
[157] P. T. Rumsby and M. M. C. Michaelis: *Phys. Lett.* **49A**, 413 (1974).
[158] T. P. Donaldson and I. J. Spalding: *Phys. Rev. Lett.* **36**, 467 (1976).
[159] G. W. Kentwell and D. A. Jones: *Phys. Reports* **145**, 319 (1987).
[160] G. B. Whitham: *Linear and Nonlinear Waves*, Wiley-Interscience, New York (1974).
[161] E. Infeld and G. Rowlands: *J. Phys.* **A12**, 2255 (1979).
[162] B. K. Shivamoggi: Unpublished (1987).
[163] V. I. Karpman and E. M. Krushkal: *Sov. Phys. JETP* **28**, 277 (1969).
[164] N. R. Pereira and L. Stenflo: *Phys. Fluids* **20**, 1733 (1977).
[165] N. R. Pereira: *Phys. Fluids* **20**, 1735 (1977).
[166] L. M. Degtyarev, R. Z. Sagdeev, G. I. Solov'ev, V. D. Shapiro, and V. I. Shevchenko: *Sov. J. Plasma Phys.* **6**, 263 (1981).
[167] H. L. Rowland, J. A. Lyon, and K. Papadopoulos: *Phys. Rev. Lett.* **36**, 346 (1981).
[168] E. Infeld and G. Rowlands: *Plasma Phys.* **19**, 343 (1977).
[169] E. W. Laedke and K. H. Spatschek: *Phys. Rev. Lett.* **39**, 1147 (1977).
[170] E. W. Laedke and K. H. Spatschek: *J. Plasma Phys.* **22**, 477 (1979).
[171] Y. V. Katyshev and V. G. Makhankov: *Phys. Lett.* **57A**, 10 (1976).
[172] E. Infeld and G. Rowlands: *J. Plasma Phys.* **19**, 343 (1977).
[173] D. Anderson, A. Bondeson, and M. Lisak: *J. Plasma Phys.* **21**, 259 (1979).
[174] M. V. Goldman, K. Rypdal, and B. Hafizi: *Phys. Fluids* **23**, 945 (1980).
[175] B. Hafizi and M. V. Goldman: *Phys. Fluids* **24**, 145 (1981).
[176] B. K. Shivamoggi: *Fizika* **18**, 205 (1986).
[177] D. Bohm: In *The Characteristics of Electrical Discharges in Magnetic Fiields*, Eds. A. Guthrie and R. W. Wakerling, McGraw-Hill, New York (1949).
[178] J. R. Cary and A. N. Kaufmann: *Phys. Fluids* **24**, 1238 (1981).
[179] R. Bingham and C. N. Lashmore-Davies: *Plasma Phys.* **21**, 433 (1979).
[180] R. Bingham and C. N. Lashmore-Davies: *J. Plasma Phys.* **21**, 51 (1979).
[181] B. N. Breizman and V. M. Malkin: *Soviet Phys. JETP* **52**, 435 (1980).
[182] J. C. Bhakta and D. Majumdar: *J. Plasma Phys.* **30**, 203 (1983).
[183] G. Pelletier: *Phys. Rev. Lett.* **49**, 782 (1982).
[184] S. Watanabe: *J. Plasma Phys.* **17**, 487 (1977).

[185] T. Honzawa and Ch. Hollenstein: *Phys. Fluids* **24**, 1806 (1981).
[186] D. A. Tidman and H. M. Stainer: *Phys. Fluids* **8**, 345 (1965).
[187] R. Hirota: *J. Phys. Soc. Japan* **33**, 1456 (1972).
[188] C. F. F. Karney, A. Sen, and F. Y. F. Chu: *Phys. Fluids* **22**, 940 (1979).
[189] C. F. F. Karney: *Phys. Fluids* **24**, 127 (1981).
[190] G. Mann and U. Motschmann: Preprint (1987).
[191] R. L. Stenzel: *Phys. Fluids* **19**, 865 (1976).
[192] K. H. Spatschek, P. K. Shukla, M. Y. Yu and V. I. Karpman: *Phys. Fluids* **22**, 576 (1979).
[193] A. C. L. Chian: *Plasma Phys.* **24**, 19 (1982).
[194] K. Nishikawa and P. K. Kaw: *Phys. Lett.* **50A**, 455 (1975).
[195] P. I. John and Y. C. Saxena: *Phys. Lett.* **56A**, 385 (1976).
[196] N. N. Rao and R. K. Varma: *Phys. Lett.* **70A**, 9 (1979).
[197] B. K. Shivamoggi: *Can. J. Phys.* **59**, 719 (1981).
[198] B. K. Shivamoggi: *J. Plasma Phys.* (1988).
[199] H. H. Kuehl: *IEEE Trans. Plasma Sci.* **PS-13**, 595 (1985).
[200] H. H. Kuehl and K. Imen: *Phys. Fluids* **28**, 2375 (1985).
[201] H. Y. Chang, S. Raychaudhuri, J. Hill, E. K. Tsikis, and K. E. Lonngren: *Phys. Fluids* **29**, 294 (1986).
[202] B. Chakraborty: *J. Math. Phys.* **11**, 2570 (1970).
[203] B. Chakraborty: *J. Math. Phys.* **12**, 529 (1971).
[204] M. N. Rosenbluth: *Phys. Rev. Lett.* **29**, 565 (1972).
[205] M. N. Rosenbluth, R. B. White, and C. S. Liu: *Phys. Rev. Lett.* **31**, 1190 (1973).
[206] D. Pesme, G. Laval, and R. Pellat: *Phys. Rev. Lett.* **31**, 203 (1973).
[207] C. S. Liu, M. N. Rosenbluth, and R. B. White: *Phys. Fluids* **17**, 1211 (1974).
[208] Y. C. Lee and P. K. Kaw: *Phys. Rev. Lett.* **32**, 135 (1974).
[209] G. C. Pramanik: *J. Plasma Phys.* **22**, 353 (1979).
[210] C. Grebogi and C. S. Liu: *J. Plasma Phys.* **23**, 147 (1980).
[211] T. Lindgren, J. Larsson, and L. Stenflo: *J. Plasma Phys.* **26**, 407 (1981).
[212] D. H. Looney and S. C. Brown: *Phys. Rev.* **93**, 965 (1954).
[213] P. J. Barrett, H. G. Jones, and R. N. Franklin: *Plasma Phys.* **10**, 911 (1968).
[214] A. Y. Wong, R. W. Motley, and N. D'Angelo: *Phys. Rev.* **133**, A436 (1964).
[215] R. L. Stenzel: *Phys. Fluids* **19**, 857 (1976).
[216] J. M. Wilcox, F. I. Boley, and A. M. DeSilva: *Phys. Fluids* **3**, 15 (1960).
[217] S. Harp: *Proc. VII International Conf. on Phenomena in Ionized Gases*, Belgrade, 1965, Vol. II, 294 (1966).
[218] K. Ko and H. Kuehl: *Phys. Rev. Lett.* **40**, 233 (1978).

Index